Bernhard Alfieri

Half-holidays with the camera

Bernhard Alfieri

Half-holidays with the camera

ISBN/EAN: 9783742827272

Manufactured in Europe, USA, Canada, Australia, Japa

Cover: Foto ©Andreas Hilbeck / pixelio.de

Manufactured and distributed by brebook publishing software
(www.brebook.com)

Bernhard Alfieri

Half-holidays with the camera

Half-Holidays with the Camera.

CHINGFORD OLD CHURCH.

HALF-HOLIDAYS

WITH THE

CAMERA.

WITH FIFTY ILLUSTRATIONS.

BY

BERNARD ALFIERI.

LONDON:

W. B. WHITTINGHAM & Co., Limited,

91, GRACECHURCH STREET, AND 43, 44 & 45, CHARTERHOUSE SQUARE.

"Our Sales are more than all the rest."

ABRIDGED PRICE LIST.

Regd. Trade Mark & Label.

ILFORD.	Plates. Ordinary and Iso. Med.	Plates. Rapid.	Plates. Special Rapid and Iso. Inst.	Papers. Bromide.	Papers. Alpha.	Bromide Opals, Per Doz.
Inches.	s. d.	s. d.	s. d.	s. d.	s. d.	s. d.
4¼ × 3¼ per dozen	1 0	1 3	1 6	0 6	0 5	1 6
5 × 4 ,,	1 7	2 0	2 6	0 9	0 8	2 6
6¼ × 3¼ ,,	2 0	2 9	3 3	..	..	..
6¾ × 4¼ ,,	2 3	2 9	3 3	..	..	..
6½ × 4¾ ,,	2 3	3 0	3 9	1 1	1 0	3 3
7½ × 5 ,,	3 5	4 6	5 3	..	..	..
8 × 5 ,,	3 9	4 9	6 0	1 5	1 4	..
8½ × 6½ ,,	4 3	5 6	6 6	1 11	1 9	..
9 × 7 ,,	5 0	6 6	8 0	..	..	..
10 × 8 ,,	7 3	9 6	10 0	2 0	2 6	10 0
12 × 10 ,,	10 6	13 0	16 0	..	..	16 0
12½ × 10½ ,,	..	..	..	4 0	3 6	..
15 × 12 ,,	18 0	23 0	24 0	..	..	24 0
15½ × 12½ ,,	..	..	..	6 6	5 6	..
18 × 15 per ½ doz.	..	..	..	5 3	..	..
20 × 16 ,,	..	..	..	5 0	5 0	..
23 × 17 ,,	..	..	..	6 9	..	..
24½ × 19 ,,	..	..	..	7 6	6 6	..
Roll. 10 ft. × 24½ in.	..	..	..	8 6	7 0	..

ILFORD P.O.P.—Pink, White or Mauve.

				With Postage.
IN SEALED TUBES.	24 Sheets	15s. 0d.	..	15s. 8d.
	12 ,, ...	7 6	..	7 11
	6 ,, ...	4 0	..	4 3
	2 ,, ...	1 4	..	1 7

Each Sheet measures 24½ in. by 17 in.

	Inches.	Per Packet.		With Postage.
IN PACKETS.	4¼ × 3¼	36 Pieces	1/-	1s. 2d.
	5 × 4	24 ,,	1/-	1 2
	6½ × 4¾	16 ,,	1/-	1 2
	8½ × 6½	9 ,,	1/-	1 3

Sole Manufacturers:—

THE BRITANNIA WORKS CO., LTD.,

ILFORD, LONDON, E.

CONTENTS.

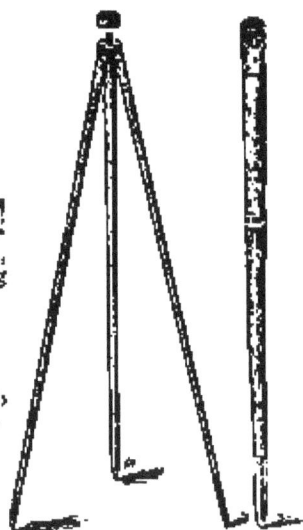

INTRODUCTORY.

— — —

NO great town is richer in suburban beauty than London,
and it is surprising to find, in spite of its rapid growth,
how soon one can get outside the radius of bricks and mortar
to the freer air beyond, and see for one's self how little the
country has really been encroached upon.

It is one of the pleasant surprises of path-finding to be
continually coming upon "Sleepy Hollow," and in the
environs of London there is such diversity of scenery as will
afford ample scope for most classes of photographers; the
pasture lands of Middlesex, the woods of Berks and Bucks,
the quiet village life of Surrey and the sea-girt shores of
Essex should all contribute to their recreation. The railways
have made a difference, certainly, and have possibly altered
the aspect of some of the country places; but it would be
uncharitable to grumble considering what facilities they give
the photographer in his search for the beautiful.

There are, however, great tracts of country still untraversed
by the iron horse; take a map and look between the Great
Eastern Railway and the Great Northern, across Hainault
(now disafforested) towards Lambourne and Ongar; and
again from Cheshunt to Hatfield, and see what a tract the
line leaves untouched. Then look at the space between the
Great Northern and the Midland, Totteridge and Shenley

way; the great gap between the North-Western and the Great Western before the Metropolitan essayed to fill it, and you will agree that there is plenty of scope for the present generation of photographers.

In this series of articles on holiday resorts round London, it will be my endeavour to give some indication of the character of the various localities from a photographic standpoint, but I hold it no part of my duty to point out exact spots from which pictures may be obtained. Who, with any claims to be considered a photographer, would thank me for such an impertinence?

Every place I have written upon I have visited, not once but many times, with the camera and without it, and I trust that the notes I have collected on these wanderings will be of real assistance to those with limited time at their disposal, for whom they were originally compiled and to whom I shall hope to prove a guide, photographer and friend.

Most of these chapters were contributed in serial form to the *Photographic Art Journal*, but "Sitting on a Gate" and one or two others are reproduced by kind permission of the editor of *Photography*.

BERNARD ALFIERI.

CRICKLEWOOD, N.W.,
   Spring, 1893.

Half-Holidays with the Camera.

CHAPTER I.

NEASDEN.—DOLLIS HILL.—KINGSBURY CHURCH AND
ITS SURROUNDINGS.—THE SLUGGISH BRENT.—
HARROW.—PINNER TO HEADSTONE.—OVER THE
FIELDS FROM PINNER TO CUCKOO HILL.—EASTCOTE
AND RUISLIP.—TO THE WOODS BY THE LANES AND
BY THE ROAD.—RUISLIP CHURCH.—THE RESERVOIR.

LET us begin with the Metropolitan Railway : it is so
essentially the Londoners' line that it will not be
considered out of place to take this enterprising Company's district first.   The Metropolitan extension, as it is
called, has recently opened out virgin country to the
photographer, and is making its way far beyond Chesham
in the direction of Aylesbury and Birmingham.

After leaving Baker Street there is nothing of interest
until you reach Willesden Green and the open country.
From here you get a view of the hog's back—Dollis Hill
—and Lord Aberdeen's house amongst the trees that
crown it ; then you speedily reach Neasden, the station
for the Welsh Harp and for one of the prettiest little
churches in the country.   This is at Kingsbury, about a
mile-and-a-half away, standing on the slopes of the lake

and thickly surrounded by fine elms. The best way to reach it after leaving Neasden Station is to turn your face northwards and follow the straight, or rather winding, road past the blocks of houses built by the Railway Company which may be called Neasden Metropolitan. You will notice the dam of the Welsh Harp on your right hand, and the sluggish Brent crossed by a wooden bridge, which at certain seasons of the year with its drooping willows makes a pretty picture (since writing this the wooden bridge has been pulled down and a new one of brick erected). Up Blackpot or Blackbird Hill you turn round by the farm house at the right, past a little pond and down a long avenue of splendid elms, until Kingsbury Church is in front of you. It is built of flint and stone, and has a wooden turret and spire of tiny dimensions, but its situation is perhaps the most striking feature. An afternoon sun suits it best, and from its difficulty of approach a wide angle lens is requisite. Blackbird farm before mentioned is a dairy farm, and the large herd of cows as they come up the lane to milking are worth consideration.

Wembley Park is the next station, but at the time of writing is not open to the public, although completed. It is here that Sir Edward Watkin proposes to build his Eiffel tower for Londoners, and the park is undergoing serious defacement to that end. There are many pretty little bits in this neighbourhood that up to the present have been difficult of access—a stream here, a farmhouse there, and tortuous lanes trending everywhere.

Harrow, photographically speaking, is not of much account, although with a long focus lens and judicious

selection you can get many pleasant glimpses of the
church on the hill, over field-paths and meadows ; but it is
a good centre to work from, and a half-hourly service of
trains from Baker Street is an additional recommendation.
It is worth while climbing the church tower for the
magnificent prospect to be obtained from the summit.
It is asserted—I do not know with what truth—that from
this outlook thirteen counties may be seen ; it may be
with the aid of a telescope and a powerful imagination, but
nevertheless, the panorama is really very striking. Northwards a long wooded range extends from Rickmansworth
and Watford to Elstree and Barnet, with the lovely country
round Stanmore, Whitchurch and Edgware intervening.
Southwards across Roxeth you can see the Surrey hills,
the Crystal Palace, Leith Hill and the Kentish Downs,
crowned by Knockholt Beeches. To the east, Epping
Forest and the Langdon hills ; and westward, the hills of
Bucks and Berks with Windsor Castle plainly visible—
altogether a prospect it would be difficult to equal from
any other standpoint round London.

Harrow Hill and its church spire seen against a
glorious sunset is a sight to be remembered, and makes a
very effective picture.

The next station is Pinner, and this village, like
Harrow, is accessible by two railways, the London and
North Western and the Metropolitan. The former is a
mile or so away, right up by the Commercial Travellers'
Schools, but the Metropolitan runs right through the
village. Pinner itself has no great attractions, but what
is worth photographing at any season—Spring, Summer,
Autumn, or Winter—is the old manor house and moat at

Headstone. Perhaps I should have mentioned before that this may be reached equally well from Harrow, although from Pinner is the more interesting walk. Going through the village to the Roxborough Lane, *i.e.*, back on the way to Harrow, you will find at the bend of the lane on the left, a rather narrow road with some fine

Headstone Manor House

elms and a copse on the right; a finger-post directs " to Headstone and Watford," which is to be followed. About half-a-mile along it the road widens out into a kind of green, and on the right hand is a gate leading up one of the broad approaches to the manor house. Through the farmyard—it is a public footpath and no one will

suggest that you are trespassing—is another gate leading
out to the road to Harrow, and from this road the best
view is to be had, as it commands the manor house, surrounded by tall ivy-grown poplars, the moat usually green
with duck-weed, and the miniature bridge connecting the
house with the farm buildings. Alas! for picture-making;
the house has been newly done up to suit the incoming
tenant, and the railings of the bridge are painted in hard

and glaring white, which shows up painfully against the
dark trees beyond.

Nevertheless it makes a very pretty picture as it
stands, and you can rarely come here without finding a
figure subject in the shape of an artist-cum-easel intent
on the same object as yourself, though by a different
method. A fair exchange is no robbery, and you can be
of use to him by supplying a copy of the photo in which
he is interested.

I am really sorry to leave Headstone, but while at Pinner there is a walk to Eastcote and Ruislip it will not do to miss. The pleasanter way in fine weather is over Cuckoo Hill and the fields; this brings you out by an old public-house with the curious title of "The Case is Altered" (there is another with the same sign at Harrow Weald), then by turning sharp round to the left, past the moated hall, you reach Eastcote, a pretty little village about two miles from Pinner. Mine host of the "Black Horse" will make you very comfortable, having a predilection for photographers, and will point out the best way to Ruislip Woods. There are two or three ways; one, and the best one, is to be found at the bottom of the village and round by the duck-pond to the right. This leads up a charming grass-grown lane and over the fields to the woods. Another is to go round by the village of Ruislip with its massive grey towered church, and then take the right-hand road again until you reach the woods from the other side.

Here you may find at Spring-time a carpet of violets and primroses; the violets (unscented) are really remarkable, and spread at intervals over many square yards. The woods have lately undergone a serious process of thinning, and there are no really big trees, but round about the reservoir there are many picturesque little bits of undergrowth and intersecting path. The public is supposed to keep to the beaten tracks, but the keepers are very indulgent with members of the craft; they seem to recognise that photography and poaching are altogether different matters.

CHAPTER II.

NORTHWOOD. — BACHER HEATH AND ITS FURZE-
COVERED COMMON.—MOOR PARK.—THE HOME
FARM, WITH ITS CATTLE STUDIES. — MILKING
TIME.—RICKMANSWORTH.—THE CANAL AND THE
COLNE. — RICKMANSWORTH PARK: FINE TREE
STUDIES. — CHORLEY WOOD COMMON: CLOUD
VIEWS. — CHALFONT: ITS ASSOCIATION WITH
MILTON.—AMERSHAM AND MISSENDEN.—CHENIES:
ITS VILLAGE GREEN AND MANOR HOUSE.—THE
RIVER CHESS.—LATIMER (LORD CHESHAM'S SEAT).
—CHESHAM AND ITS INDUSTRY.

THE last station we got out at was Pinner; the next
is Northwood, a new neighbourhood rapidly
accumulating bricks and mortar. There is nothing of
interest to be seen from the line, if we except a distant
view of Ruislip Reservoir lying in the surrounding woods.
Northwood is in fact the best station for it, and the walk
there is charming. Seeing where the reservoir lies from
the railway carriage, one can scarcely miss it. Coming
from the station by the new road you turn to the left on
reaching the main road, and where the road forks further
on taking the right again, which terminates in a field-path.
It will be well to notice on the way a delightful bit of
common land with a sandy road across it, its shallow pools
and gravel ridges losing themselves in a wealth of bracken.
Reaching the field-path you pass by a farm-yard and soon

arrive at the reservoir—at this end a very charming piece of water which is strictly preserved for fishing and shooting.

If, on the other hand, on reaching the main road from Northwood Station you turn to the right, you will find a

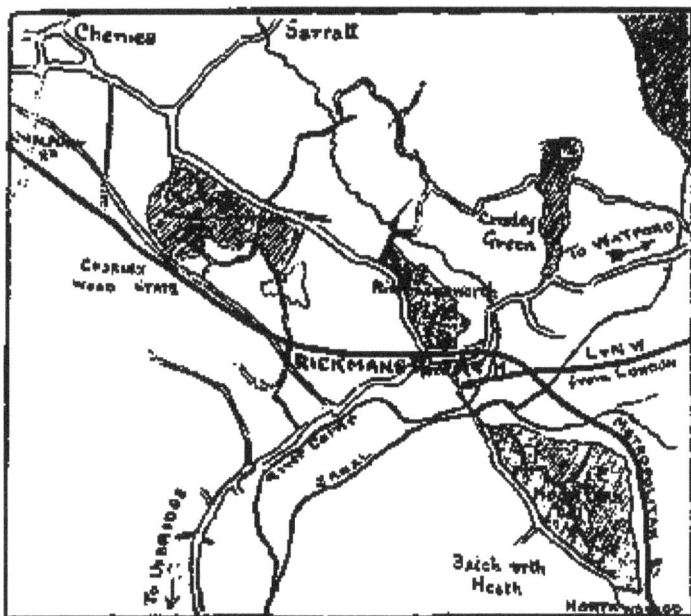

long and steady rise of about a mile to a furze-covered common studded with miniature ponds and picturesque red-brick houses, known as Bacher Heath or Batchworth Heath—apparently nomenclature is of no consequence in this neighbourhood.

Facing the common—on which, by the way, an

occasional gipsy encampment is to be met with, as well as a most picturesque village pound, green with age and disuse, and almost buried in briar and bracken—is one of the entrances to Moor Park, a massive stone gateway, the seat of Lord Ebury. The public entrance is beyond the last house on the common—a cottage that strongly calls to mind Birket Foster—by a wicket gate on the right hand; a finger-post indicates a public path to Rickmansworth. This traverses Moor Park, and by following it one is sure to come across studies of cattle, deer and old trees. On the left of the path, lying in the hollow, is the home farm, or one of them at least, with a pond close by. It is interesting when at work in the park to note what regular timekeepers the cows are. In summer time at five minutes to four you can see them with one accord trooping down to the farm gate to be milked. They usually have a minute or two to wait until the boy comes to let them in, and while they are waiting is a favourable opportunity for the photographer and his cattle studies. It is a very accommodating boy, too, and with judicious management he may be induced to delay opening the gate for a little time, to allow the photographer to finish his work.

There is a fine herd of fallow deer also, and altogether the man of snap-shots ought to have a happy time. He whose snap-shots are premeditated will be able to do good work.

With the exception of a few fine old oaks—in good condition and otherwise—there is not much else to do in Moor Park. The mansion is in the Italian style, and dreadfully modern-looking and conventional (if his lordship will excuse me saying so), but it is very pleasant to be allowed to take this way to Rickmansworth, instead of having to go round by the road; and the regulations as to keeping to the footpaths are not, I think, enforced with undue rigour.

At the other end of the park Rickmansworth is reached. The Canal and the river Colne have some picturesque bits to be found after diligent search. The church is a fine old crusted specimen, which has lately undergone restoration, and the rest of the town is of the straggling kind, boasting many public-houses.

As far as one can make out, there is nothing surprisingly picturesque to be found in it. It possesses two railway stations—one on the Metropolitan, at the further end of the town, and the other on the London and North-Western, by a branch line from Watford, opposite the church. Rickmansworth Park (altogether different from Moor Park) is very well worth visiting for tree studies, curious as well as varied, and is reached from the Metropolitan station very readily by the bridge crossing the line.

The ways out of Rickmansworth to Harefield and Denham by the Canal, and to Cashiobury Park, either by the canal or the road, I shall hope to deal with in another chapter, and for the present we will get on to Chorley Wood, or Charley Wood as the natives prefer to call it, the next station beyond Rickmansworth, on the new extension to Chesham. The Common is a bare and bleak plateau, barren of interest except to the cloud-catcher. Here he can fix his tripod and command the heavens at all points of the compass, uninterrupted by trees or houses. I remember, after an unsuccessful day

at Chesham in the autumn of last year, returning by rail
with a friend in the evening, we saw on arrival at Chorley
Wood some fine sunset effects from the carriage windows,
which compelled us to get out there—although there
was not another train for two hours—to secure them.
Chalfont Road comes next, and is the centre for Chalfont
St. Giles and Chalfont St. Peter—historical from their

Amersham from the fields.

connection with Milton and " Paradise Lost "—while an
omnibus service connects it with Amersham and Missenden. The former is another of the straggling kind, old-
fashioned and respectable, with nothing to commend it
to the photographer except a field-path on the other side
of the stream, round by the mill and over the cornfields,
where a bird's-eye view of the town is to be had from the
rising ground. With the shocks of corn in the foreground,

this view is by no means to be despised, and they have a way of binding up the sheaves which fits in with the photographer's idea of the way it should be done,

Chenies is more varied, and lies north of Chalfont Road Station. By following the road a little distance you will see a path over the fields, and at the end of it the old manor house of Chenies. There is in front of it —the back of the house, by the way—a historical oak tree which helps to make a picture. For this a morning light is best, and for the front of the house one would require to be there early in the day.

The village green is planted with fine elms, and is such a one as is rarely seen, while at the bottom of the road lies the valley of the winding Chess, a jealously-guarded trout stream, crossed by a path that leads to Sarratt and Sarratt Bottom.

I say jealously-guarded of a set purpose, for the old keeper here is a very dragon. You may be studying the river from a pictorial point of view when he comes suddenly upon you and concludes it piscatorial—unlawfully piscatorial. Drink of the water in the river and he imagines you are swallowing the fish : but, after all, he improves upon acquaintance, and is peculiarly susceptible to the blandishments of the photographer. May his shadow never grow less.

Following the road you pass by stately Latimer, standing on a commanding eminence at the other side of the river, the seat of Lord Chesham, until the town of the latter name is reached.

Chesham is the home of the wood industry; wooden bowls, wooden shoes, everything that is wooden comes from Chesham. The River Chess opens out in the middle of the town, having the appearance of a miniature lake, and at certain points, which are far from frequent, a picture may be obtained. By the joint enterprise of the inhabitants and the Metropolitan Railway, the town has recently been connected with the outside world. Until this desirable event happened there was no railway within many miles, and trade was severely handicapped. Chesham was, until recently, the terminus of the Company, but the line has already been pushed considerably beyond it, but in a more Westerly direction. Amersham, Missenden and Wendover have stations of their own, and the present terminus, a temporary one, is at Aylesbury.

All this is new country to the photographer, and Wendover lying at the foot of the Chiltern Hills, is especially charming. From Wendover there runs an old canal ; all along it the scenery is very pretty, and one or two of its quiet reaches are worth photographing. It passes in front of Halton, Mr. Alfred Rothchild's house, leaves Wilstone reservoir down in the hollow, and finally joins the Grand Junction at Tring.

It will not, of course, be attempted to bring the whole of the places enumerated in this chapter into one day's work; but, given an idea of the various features of each of the centres on this line of railway beyond Pinner, the photographer will be able to make a selection according to his needs. For cloud studies Chorley Wood Common is equal, if not superior, to Parliament Hill and Hampstead Heath, as one escapes the smoky atmosphere of the great city.

CHAPTER III.

WEST DRAYTON, RIVER COLNE, IVER

To West Drayton by the Great Western Railway.—The River Colne and its Many Branches.—Drayton Mill; a Long Wooden Bridge and River Scenery.—West Drayton Church and the Old "Gate House."—The "Trout" Inn; a Stretch of Meadow Land.—On the Canal and the Road to Iver.

WEST DRAYTON is situated on the Great Western Railway, and is thirteen miles from London. Viewed from the line it is about as unpromising a photographic centre as can well be imagined; the whole of the surrounding country is terribly flat; brickfields everywhere, and the houses apparently of one design, dull and uninteresting. There is only one circumstance that makes West Drayton worth visiting for photographic purposes, and it is that the river Colne splits up into several branches and waters it abundantly. It seems, though, that the Colne being a trout stream, and naturally a "subscription" water, is preserved very severely, and the right of way is denied to other than fishermen on both sides of its banks. The exceptions to this rule make one wish that it were not so, and there is scarcely a stretch of river open to the public which is not worth the photographer's attention.

The little sketch map below gives a diversity of roads
round about West Drayton, and there are many cross
roads and bye-lanes I have omitted. Many of these
little lanes are very charming and should yield good

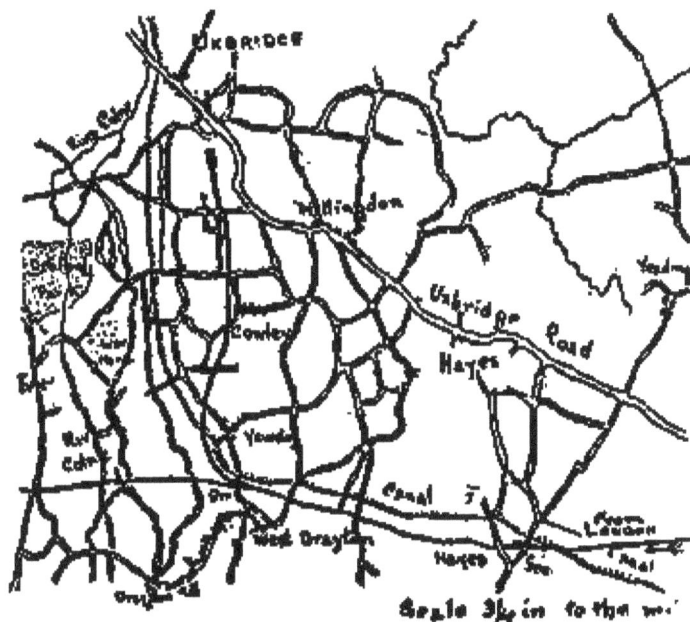

Scale 3¼ in. to the mile

results if the photographer has the time and leisure to
leave the railway any distance in his rear, but for the
present there are one or two points in its immediate
vicinity it will be well to visit.

Coming from the railway station you turn under the
bridge and follow the road bounded by a wall on its right
until you reach the village green. Notice the fine elms
on the other side of it, together with a picturesque

cottage or two, and a substantial red brick farmhouse at
its further extremity. Past the green the road ends at
Drayton Mill. There is an inn called the "Angler's
Arms" on one side, and a pretty three-arched brick
bridge on the other. One branch of the Colne comes
swirling under the arches, while another finds its way
under the huge mill, swallowing up the road in its onward
rush. This surprises you for a moment or two, but when
you look more closely you see cart tracks in the gravel
bed and find that the river is fordable.

This is a point where lingering is permissible in case
a market cart turns up to help your picture-making, but
let us get back to the "Angler's Arms." A footpath here
accompanies the river for some distance, and almost
every yard of it should be considered; the water is very
clear, but in summer time it is well broken up with white
blossom (not the arrow-head, but another water-weed)
and green rushes, while the piles driven in here and there,
bringing up the masses of drifting weed, break it still
further.

On the other side of the river is a line of pollarded
willows and a pretty ivy-grown cottage, and looking back
towards the bridge, you see over the tree tops the long
high roof of Drayton Mill. It is too bare in winter from
this point, but in summer time much of the outline is
concealed by foliage. Further down the path a long
wooden bridge crosses the river where the other branch
joins it, and there is again some distance open to the
photographer. I have seen this stretch of water at all
seasons of the year, and the pictures to be found upon it
are many; it seems to be too quickly running to freeze

entirely, and in the severest frost you will find the channels free, although the moor-hens are disconsolately grouped together on the over-hanging ice-floes.

Before leaving West Drayton for another bit of the Colne where right of way is not denied, it will be well to have a look at the church, best reached by the road that skirts the green. It is a solid-looking structure, built of flints, but with a turret of stone, brick and wood combined; the tower is completely covered with ivy, and from the road the whole building is partially concealed by a high wall. Close by the entrance to the church-yard is a low red-brick gate-house, with a pointed arch, and a turret on either side. This is all, or nearly all, that remains of the manor house of West Drayton, and its best feature is, I think, the oak door, which has every appearance of great antiquity.

Retracing your steps to the railway station, turn to the left under the bridge: it will bring you to another branch of the river after five or six minutes' walk. Here is the "Trout" Inn, and a stretch of four or five meadows adjoins it, through which runs a footpath until abruptly brought to an end by an iron bridge. This bridge carries over the canal, and it is a matter of no difficulty to gain the towing path; turn now to the right, and reaching the junction of the other canal, cross the bridge and take the left-hand path in the direction of Uxbridge. At this point there is nothing of interest whatever, but over the next stone bridge is the road to Iver, and by turning to the left you come across the river again. The first arm of it is open for a little distance only, but there is a group of cottages with a long line of willows stretching out towards them some little way up the bank.

In the last Photographic Exhibition, this subject
with a group of children in the foreground made a very
effective picture, and more than one standpoint is to be
obtained. This, however, is not such a pretty branch of
the river as one that you will find two or three hundred
yards further down the road; here is a ford and a pretty

On the Colne at Iver.

wooden bridge with a charming background of foliage,
while the swaying reeds and water-plants in the clear
river are frequently assisted in making a picture by the
trout angler on the bank. It is this bridge that forms the
subject of the sketch.

I am afraid it is not much use following the road
further down towards Iver, although about half a mile
beyond, the Colne crosses again, this time winding round

by the church and village through pleasant meadow land;
it necessitates further walking than the average photographer is prepared to undertake, and he would probably
come to the conclusion that the view was of no great
value.

West Drayton is worth a visit, but the disadvantage
of not being allowed free access to the river Colne is, to
say the least of it, very annoying.

CHAPTER IV.

BURNHAM BEECHES.

BURNHAM BEECHES AND HOW TO GET THERE: COACH
FROM SLOUGH THROUGH FARNHAM ROYAL TO EAST
BURNHAM. — A TINY STREAM. — THE "NODDING
BEECH" OF GRAY'S "ELEGY." — STOKE POGES
CHURCH. — TREE STUDIES AND RUSH-GROWN POOLS.
— TWO NOTABLE STUMPS ON THE WESTERN SIDE
OF THE "BEECHES." — AS TO EXPOSURE AND THE
BEST SEASON FOR PHOTOGRAPHIC WORK.

IF anyone asked me to indicate a better place for photographic work near London than Burnham Beeches, I should have to acknowledge at once my inability to do it; for tree studies of varied kinds, for wild foregrounds of briar and bracken, and for two of the most charming rushy ponds I know, there is no more favourable locality to be found. Burnham Beeches became public property in 1879, through the energy of the City authorities prompted thereto mainly by Mr. Francis Geo. Heath, brother of the Vernon Heath whose photographs have brought it into prominence. It is a tract of wild woodland situated between Slough on the one hand, and Beaconsfield on the other, its nearest communication by railway being from the first-named station on the Great Western Railway. All through the summer until late autumn the Company offers special inducements to visitors in the way of cheap fares and a quick train service, issuing return tickets including coach from Slough at 4s.

Passengers who book through to the Beeches take precedence on the coach over local visitors, and although this may seem on the face of it an unnecessary statement,

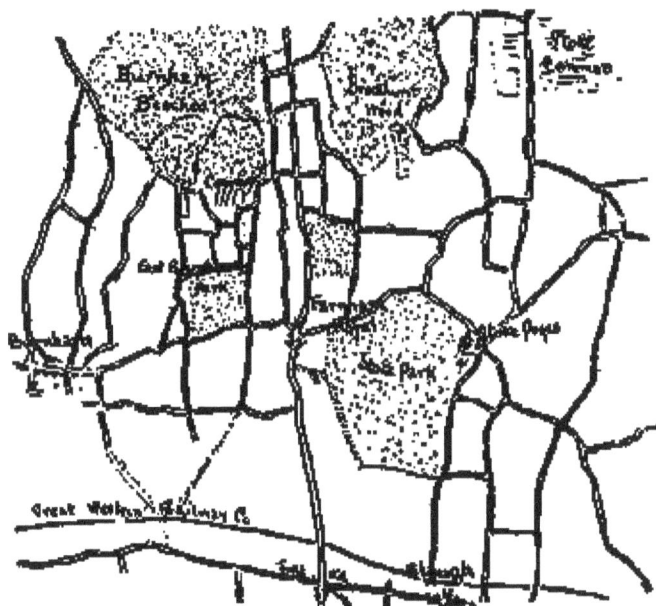

I may add that on Bank Holidays and special occasions I have seen every coach filled up immediately on its arrival in the station yard, and every ticket subjected to scrutiny, that those who have travelled the longest distance may have the preference. Generally speaking, however, there will be no difficulty in obtaining a seat, and the only day on which I would recommend photographers to avoid Burnham Beeches is the August Bank Holiday, for not only does it recall Epping Forest at a similar time, but the " Beeches " are full of people and donkeys, and although animal studies are very well in their way, they do not always harmonise with the sylvan scenery you have come to photograph.

After this long preamble it will be well to proceed. The coach ride is delightful except on a very hot dusty day—not altogether rare on these sandy roads—and at almost every turn you come across something worth studying. Notice what a beautiful view you get of Windsor Castle on the other side of the river as the road turns abruptly to the right on the way to Farnham Royal, the distant glimpse of Clieveden Woods beyond Maidenhead, and the nearer view of Stoke Park a little farther up the road you are travelling after crossing the railway bridge. Farnham Royal is about three miles by the ordnance survey, and after leaving the church on the right the coach turns the other way through the village, and on to East Burnham; you pass East Burnham Park, once inhabited by Richard Brinsley Sheridan, and then of two roads leading to the Beeches the coach takes the one at the left. That on the right passes the " Crown " (the only public house as far as I know in the vicinity), and if you are on foot is equally as convenient as the other. Following the coach, you shortly reach a tiny brook shaded by a magnificent tree, which Mr. Heath, in his work on Burnham Beeches, indicates as the "nodding beech " of Gray's " Elegy." The mention of this leads me to say that Stoke Poges Church does not lie in the route of the coach, which is to be regretted, as it makes a very presentable picture in spite of the new fence they have put round the ancient graveyard.

Once past the boundary formed by the brook there is such a wealth of photographic subjects that the camera

In Burnham Beeches.

man will have some difficulty in making a selection ; it seems that, put your tripod where you will, you may almost take shots at random with satisfactory results. Every road through the " Beeches " is full of tree-studies ; every bye-path a picture with its tangled undergrowth and mossy velvet carpet—the gnarled trunks on every hand disputing passage and the gravelly ridges and shady

dells affording work not for one day but for many. There
are two pools in Burnham Beeches, and charming pools
they are; the one nearest the Common—that is, on the
east side—is rather smaller than the one on the south,
but you will find upon it as a rule one or two huts
tenanted by enthusiastic painters, whose delight it is to
transfer to canvas the changing moods of nature
reflected on the rushy shallows. Both these pools have
characteristic backgrounds, but the larger one is rather
difficult for photographic purposes on account of its
being so shut in by surrounding woodland. At the west
of the "Beeches," on the other side of "Seven-ways
Plain," are two grand stumps of fallen monarchs that
have battled for ages against wind and weather until they
can hold their own no longer; and the living shroud of
green ferns that encircles their picturesque forms only
adds to their majestic beauty.

A word as to exposure may be permitted me: in
summer and autumn many of the beautiful glades in
Burnham Beeches are so shaded by heavy foliage that a
long exposure is necessary to give detail. The tendency
is—I speak from my own experience—to err on the
wrong side in not giving sufficient time when using a
small-top. This may be easily avoided. Halation, too,
frequently occurs unless the necessary precautions are
taken to obviate it, as they should be in work of this
nature; even a thickly-coated plate is none the worse for
"backing" when photographing tree subjects where the
light glimmers through the foliage in unexpected places.

The best season of the year for photographic work in
Burnham Beeches I take to be autumn, for then you

have the bracken in its full beauty—some of it growing six feet high, too—and you can generally reckon on the absence of wind, which in the earlier summer does so much to mar tree pictures. Winter has its charms also, and where can you get such studies of hoar-frost as here? Burnham at all seasons is beautiful, and though you may have to walk from Slough in visiting it while the coach service is suspended, what does it matter when you can cut off nearly a mile by the field paths, and take in Stoke Poges by the way?

CHAPTER V.

PERIVALE, EALING, GREENFORD, AND NORTHOLT.

ON THE PADDINGTON CANAL.—TWYFORD ABBEY.—
HANGER HILL.—THE RIVER BRENT.—PERIVALE:
ITS CURIOUS CHURCH.—GREENFORD AND NORTHOLT.

TO look at Willesden from the railway with its network
of lines in all directions, affords at first sight no lively
prospect for the photographer's work. Yet Willesden is
by no means such a bad centre after all; the canal,
uninteresting enough just here, soon assumes a prettier
aspect; the river Brent wanders through the grounds of
Twyford Abbey, and from there on to Perivale and
Greenford. But to proceed in geographical order. After
leaving Willesden Station—and it is difficult to leave
from the high level owing to the fact that you have to
make your way to the up-platform of the main line
(London and North-Western) and find your way out to
the road from there—you turn to the left and cross the
bridge; then about a quarter of a mile down the road
you will see the works on the canal where "Willesden
paper" is made; they are easily distinguishable, from the
tall brick chimney which rises above them. Over the
canal bridge you can descend to the towing-path. Follow
it until, after passing all signs of habitation, you reach a
bridge which joins a road running parallel with the

towing-path. This road must be taken; it leads through
a farmyard (constituting nearly the whole parish of
Twyford), and becomes a private road to Twyford Abbey.
It is rather a handsome house—secular, of course—with
octagonal turrets and many windows. Standing on the

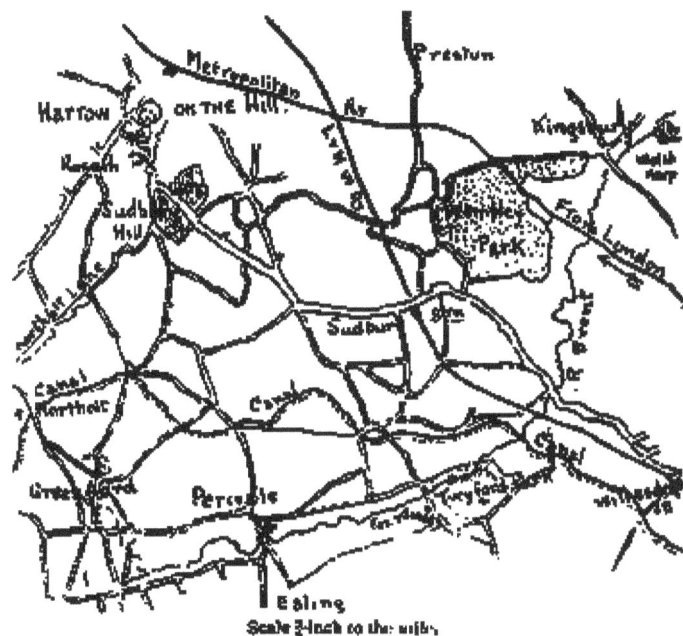

Scale ¾-Inch to the mile,

lawn within a stone's throw of the house, covered with
masses of ivy, is Twyford Church, and a notice at the
lodge gates informs you of the hours of service. Presumably the park is to be considered private, but most
probably no difficulty would be met with in obtaining
permission to photograph.

The road past the Abbey gates is now an avenue with broad grassy borders, and presently emerges through the park gates to Hanger Hill. By turning to the right past the "Fox and Goose," and down the hill, you come upon a little bridge crossing the Brent, with a cluster of cottages at the waterside on the right. A finger-post directs by a road immediately at the left of the bridge to Perivale and Greenford. Roughly speaking, it is an hour's walk to Perivale from Willesden by this route, and going round by the canal to Alperton, and from there to Perivale, would take about the same time; altogether a nearer way would be by taking train (Great Western) to Ealing, and walking from there to this point or to Sudbury (London and North-Western Railway) and coming through Alperton, but then it misses Twyford Abbey.

Perivale Church and rectory are so curious that a few words of description will not be out of place; the chancel and nave are either of stone or brick, but entirely covered with plaster, while the little tower surmounted by a cap is of wood. Let in the woodwork are two or three little windows, and a sun-dial occupies a place at the side, about half-way up. The porch is very pretty, and covered with ivy, and, taken altogether, the church with its red tiles, its surrounding trees overshadowing the little graveyard, forms a very pretty picture worth securing, though the whole of it is of such small dimensions that it looks like a toy or the creation of a passing fancy. The rectory adjoining fits in with the whim; in style a kind of Elizabethan, wood and plaster, and, to complete the eccentricity of a parish which, all told, numbers twenty-one souls, the rectory porch has a case of ornithological specimens—a couple of flamingoes amongst them—as its chief ornament.

For photographic purposes a morning light best suits the south side of the church, but it is difficult to get at

Perivale Church

from that side, and later on in the afternoon is the best time for the rectory and church combined.

The footpath to the church is continued by a long wooden bridge over the river Brent to Ealing, and although there is nothing very striking to be found on its banks, studies of cattle at the brookside may be met with. I make no apology for mentioning this stretch of pleasant meadow land, with the stream winding through it, for, although many amateurs would not think of exposing a plate on it, to those who can make a satisfactory picture of a haystack and a five-barred gate by judicious arrangement, I think there is sufficient inducement offered to pay it a visit.

It is not a very far cry from Perivale to Greenford by the fields, or, for the matter of that, by the road. In this quiet village you come across the Brent again; there are many little interesting bits by its side. Northolt, too, is not far away, but a glance at the map will show that a different direction must be taken.

Northolt Church is the distinguishing feature of the little hamlet, which is built on two sides of a miniature valley ; it stands on an eminence overlooking the cluster of cottages claiming its protection. Here is another curious specimen of church architecture, built of brick and plaster, and flanked with many buttresses, a tiny little building with a wooden steeple, and a roof of weather-beaten, many-coloured tiles. The porch is of wood, and covered with ivy, and, like the rest of the church, bears unmistakable signs of age. Harrow is the nearest station from Northolt, and it may be said at once that from Willesden to Twyford Abbey, Perivale, Greenford, and Northolt is a tolerably long stretch of nine or ten miles, but some of this may be curtailed by leaving Willesden out of calculation altogether, and taking train to Ealing. The chief interest on the whole round undoubtedly centres in the wooden spire of Perivale and its neighbouring rectory. At the same

time a variation may be made by following the banks of
the canal from Alperton. On the London side particularly there is a very pretty bit of landscape, where
the Brent tunnels under the canal, and makes its appearance in the grounds of Twyford Abbey. In the shade of
many trees the river is almost secluded, and the king-
fisher is frequently to be seen.

CHAPTER VI.

HENDON, THE HALE, WHITCHURCH AND STANMORE.

BY THE MIDLAND TO MILL HILL.—HENDON CHURCH,
AND THE VIEW FROM IT.—A SEQUESTERED HAMLET:
THE HALE.—OVER THE FIELDS TO EDGWARE,
"CANONS" AND WHITCHURCH.—HANDEL'S ORGAN.
—"THE HARMONIOUS BLACKSMITH."—A DIFFICULT
CHURCH TO PHOTOGRAPH.—FIELD-PATH ACROSS
"CANONS" PARK AND THE AVENUE.—STANMORE
CHURCH, "ANCIENT AND MODERN": A PICTURESQUE
RUIN.—BACK ACROSS THE FIELDS TO EDGWARE.—
A FORGOTTEN CORNER OF THE COMMON, AND A
PRETTY LANE.

THE Midland Railway in the immediate neighbourhood
of London passes through no specially interesting
country until Mill Hill is left behind. I purposely pass
over Hendon, because, except for being a convenient
station for Kingsbury Church, mentioned in a previous
chapter, there is nothing of photographic value in the
vicinity. Hendon Church is finely situated, and its
churchyard, with an avenue of clipped limes, and its yew
trees, is prettier than the average, but its chief charm is
the extensive view it commands over the uplands of
Stanmore, Bushey Heath, and the Buckinghamshire
hills beyond.

Mill Hill, the next station, is better; convenient for Highwood Hill and Totteridge Common, as it is nearest on this line of railway for Edgware. It is a very pretty walk from Mill Hill to a little cluster of houses known as " The Hale," then past the " Green Man " (our good

old friend Bates is there no longer), and over the fields until you reach the road close by Edgware Church. It is, however, a saving in time to take the Great Northern branch line, via Finsbury Park, to Edgware itself, as the station is right in the centre of the village. An old-fashioned village it is, too, consisting of one long rambling

street—half of which, by the way, belongs to another
parish; it has that forlorn look upon it which is a
standing reproach to the railway "innovation" and the
decadence of coaching.

Coming from the station, the road crosses what is
supposed to be the ancient Watling Street, then it skirts
Canons Park, formerly the seat of the first Duke of
Buckingham and Chandos, and in about half a mile
Whitchurch is reached. This church is worth more
than a cursory inspection on account of its association
with Handel, who officiated here for some years as
organist while under the "patronage" of the man
whom Pope and Dean Swift satirised so bitterly. I
will not weary you with historical reminiscences, but
the photographer ought to know something about a place
which is so full of them. Sir James Brydges was in
Queen Anne's time paymaster of the forces, and by
virtue of his talents, or his office, accumulated an immense fortune. With this he built Canons and the
church in connection with it, at a cost estimated at a
quarter of a million. Pope's prophecy was fulfilled in
less time than he anticipated, and "the wonder of the
age" was pulled down after a brief span of thirty years.
Another "villa" of more modest dimensions now reigns
in its stead.

To come to modern times. You will have no difficulty
in obtaining the keys of Whitchurch from the rectory;
a charge of sixpence is made—proceeds devoted to some
worthy object or other—and the privilege is worth it.
Notice close to the gravel path a stone which has replaced
the wooden rail, dedicated to the memory of William

Powell, the "Harmonious Blacksmith." You know the
story, but the forge where Handel took refuge in the
storm is to be seen no longer.

Directly you get inside the church, the lavish display

Tower of Stanmore old Church

of gold and colour—mellowed with age—the stained
windows and the mural decorations, will strike you at
once. On the ceilings, on the walls at each side of the
altar, are the masterpieces of Belucci, while perched up
aloft, behind the screen at the east end, is the veritable
organ on which Handel played, still fit and used for
service. In a side chapel the "monument room," as it
is called, contains a memorial in marble to the first duke.
There is here, in fact, everything to make the church

worth a visit; but photography, owing to the stained windows and deep local colouring, is difficult, and a very long exposure would be necessary.

Leaving the church you will find, a few yards up the road, a footpath on the right, striking across Canons Park, now cut up for agricultural purposes; it passes a pond with a clump of trees in its centre, and then traverses the avenue leading up to the modern house of "Canons"; then over more fields in which groups of cattle are usually to be found, until it rejoins the road you left at Whitchurch, within a stone's throw almost of aristocratic Stanmore. Here a choice of routes is open. In the centre of the village the road on the right leads up Stanmore Hill—a very steep one—and brings you to Bushey Heath and Stanmore Common; but by keeping straight on you reach the church, ancient and modern—a new and substantial one, built in the graveyard consecrated to the old. The beautiful ruins are still standing; the fine massive red-brick tower, almost completely covered with clustering ivy, is the home of the bats and the screech-owl, and as fine a subject for "halation" as you shall find in a day's march.

It makes a picture in spite of this disadvantage, and the added difficulty of approach. The south side seems best, as it is more open, and an early afternoon light brings the dark masses of ivy into relief. It is not an easy subject at any time, but it is worth some expenditure of trouble to secure.

Within a few hundred yards south of the church a new station has quiet recently been opened connecting Stanmore with the London and North Western main line

at Harrow. At present there are about nine trains a day
each way, except Sundays, but it is really not far to walk
to Harrow, passing on the way a very primitive little inn,
with a pond in front of it, known as the "Duck in the Pond."
If you prefer to return to Edgware, take the road past
the station, then turn across by the first footpath over
the fields to the left. This dives under the railway and
joins a pretty winding lane which comes out on a bit of
common land almost in front of the avenue up to
"Canons," and the gasworks supplying Edgware and
Stanmore. The gasometer does not make a picture, but
there are some little rush-grown ponds dotted about on
the common which are worth looking at. In the far
corner, apparently forgotten by the world, is another
"Green Man." The road in front of it gradually narrows
as it winds, getting wilder and more deeply rutted every
few yards : just one of those grassy lanes that artists
and botanists like to come across, but provocative of
strong language from a driver's point of view—a lane
of many turnings, finally emerging on the road that
connects Kingsbury with Harrow.

CHAPTER VII.

CASHIOBURY PARK.

WATFORD TO CASHIOBURY PARK.—CASHIO BRIDGE.—
THE HALFWAY HOUSE.—STUDIES OF GEESE.—THE
CANAL AND THE RIVER.—MILL AND LOCKS.—CANAL
LIFE.

I HAVE given no map on this journey, as the road is
quite plain and straightforward from Watford, and
the one from Rickmansworth is hardly less so.

Cashiobury Park is nearer to Watford by a mile or so,
but nevertheless, I prefer on most occasions walking from
Rickmansworth. Perhaps it is that there is a regular
service of trains on the Metropolitan, and, remembering
the number of minutes past every hour at which they
start, one never need fidget with a time-table. Besides,
the walk by the canal into the park is too pleasant to
miss for the sake of an extra mile saved in distance.

Coming from Watford Station you make your way to
the main road and turn to the left in the direction of
Rickmansworth, crossing in less than half a mile another
road which leads from Watford to King's Langley and
Boxmoor. You are by this time well outside the town,
and a couple of hundred yards or so down the road the
gateway of Cashiobury Park stares you in the face. This
is, perhaps, the best way in for those whose time is limited,
but if not, it is preferable to keep down the lane skirting

the park for half a mile until you reach Cashio Bridge, a
point I will deal with presently.

If you take Rickmansworth as your starting point,
there are three ways. The first is viá Croxley Green

OLD LOCK, CASHIOBURY PARK.

and needs no comment; the second is prettier and
requires some directions to enable you to find it. Turn
under the bridge from the railway station and immediately to the left you will notice a flight of steps leading
up to a path running parallel with the railway; this

saves something in distance, for you join the Watford road by the third side of a triangle. The brewery is opposite you at this point, and at the side of it is a footpath which will bring you eventually to the road, from which you can see Cashiobury Park a little distance further on. The third is by the canal, which you reach by going through the town beyond the church, and turning to the left after having gained the towing-path—the pleasantest way of the three if you have the time to take it. All these routes converge at Cashio bridge, where there is an inn called the " Half-way House." It is not a pretentious-looking place, but it is clean and comfortable ; and the landlord, a very worthy and obliging fellow, will attend to your wants with alacrity and subsequent moderation in the way of charge. I mention this inn particularly because it is a recognised resting-place for the bargees, with a wharf attached to it ; and on Saturday afternoons after having knocked-off work for the week you can generally rely on a string of barge horses being brought to water. The river is pretty, the background also, and a more genial set than the bargees it has rarely been my luck to come across. I have spent many pleasant afternoons with them.

The stream running by the side of the inn is the river Gade, hurrying to join the Colne lower down at Rickmansworth ; it is a trout stream of the first order, and preserved, as far as fishing is concerned, religiously ; but you will find a path by the side of it carried across by a pretty little wooden bridge, past some water-cress beds to the canal bridge, on the other side of which you can get into the park by negotiating a fence. The more direct way from Cashio Bridge to this point is a course by the canal towing-path, but in taking this way you miss a pretty little bit of river scenery and marsh land, a favourite resort of all the geese in the neighbourhood.

Having got into the park, a notice board will inform you that wilful damage of any sort is looked upon with an unfavourable eye, but otherwise the place is, by the kindness of its proprietor, the Earl of Essex, at the service of the public, and there need be no fear of trespassing. The canal runs right through this beautiful domain, and so does the river Gade ; on the grassy slopes are many fine tree studies; dotted here and there are curious specimens of long-haired and horned sheep, herds of deer, and cattle studies of all kinds innumerable. The deer, however, are rather shy, and require stalking with a long-focus lens to ensure good sport.

At the first lock there is a bridge, over which runs the road coming from the entrance we noticed after leaving Watford, and standing in a fork of the river is a mill of rather modern construction, but picturesque enough from certain standpoints. It is surrounded by fine elms, oaks and hornbeams, and a waterfall may be secured on the one side of it when the river is sufficiently high. Altogether it is a charming spot on a summer's day; you can see the trout in dozens together, and when the "May fly" is on they make the water alive with their splashes. A morning light is best for the mill, with the waterfall from the roadway, but an afternoon for the best point of view to be had round by the lock and its adjacent fence.

It is never worth while hurrying away from this lock

D

when you are seeking to illustrate canal life; fresh barges
are always coming along, and as it takes some time to
get through the lock, there is no necessity to hurry yourself and expose the same plate twice, or anything of that
sort. If the grouping does not agree with your ideas
of composition, or the horse is of a wrong colour, it is
well to wait awhile until the right one turns up. A
quarter of a mile away is a second lock, prefaced by a
wooden bridge under which comes the rapid-running
Gade to join the waters of the canal. I want to point
out this lock to you, because, towering above it is a
magnificent chestnut, which, when in full bloom, is a
sight worth seeing; it is a good shape, moreover (photographically), and helps wonderfully in the composition
of the picture.

A little further on there is still another lock, but it is
rather bare and uninteresting in spite of a Scotch fir
that overhangs it; and the canal becomes just an ordinary
canal for half a mile or so. Then you cross it by a
bridge, and its character changes all at once, and with
massive elms and ivy-covered oaks presents to your
notice as fine a bit of river scenery as you shall find in
the home counties. With a barge in the foreground or
the middle distance to break it up, this is certainly a
picture, but I would strongly recommend thickly-coated
plates or "backing" to avoid what usually occurs at this
point—halation. Grove Park lies at the other end of
this stretch of canal, and a stone bridge connects it with
the towing-path. I mention this bridge because you may
frequently see a beautiful herd of Jersey cattle crossing
it to find fresh pastures. The park itself is private.

The canal may be followed, if you are in the humour to follow, up to Hunton Bridge and King's Langley, and, for the matter of that, very much further; it is a succession of twists and turns, and many locks which may claim the photographer's attention. At Lady Capel's wharf there is a bridge, and by the road over it you can join the main road back to Watford; it saves a little in time, but it is a hot and dusty road in summer and the towing-path through the park again is preferable.

No permission to photograph is needed, and there is sufficient work in Cashiobury Park for a whole day—a very pleasant day, too; its attractions are varied, and it is quiet and secluded at all times.

CHAPTER VIII.

WATFORD, ALDENHAM, RADLETT, OR ELSTREE

LONDON TO WATFORD AND BUSHEY.—VIEWS ON THE
HEATH.—BUSHEY MILL.—ALDENHAM: THE CHURCH
AND SOME HAYSTACKS.—"THE CHEQUERS," AND A
FOOTPATH TO ALDENHAM ABBEY.— THE RIVER
COLNE: A WOODEN BRIDGE AND A WATERFALL.—TO
GREAT OTTER'S POOL.—THE WAYS OF HIGHLAND
AND OTHER CATTLE.— LETCHMOOR HEATH AND
PATCHETT'S GREEN.—HOME BY WAY OF RADLETT
OR ELSTREE.

WATFORD is situated on the L. and N. W. Railway,
fourteen miles from London, and besides being the
station for Cashiobury Park, it is a good point from which
to reach Aldenham, Patchett's Green, and Letchmoor
Heath.   Bushey, however, serves equally well and is a
station nearer London, but while talking of Bushey, it
should be mentioned that the Heath—also called Stanmore Common—is two miles south, and is better reached
from the new station at Stanmore.   For the particular
class of views that a heath affords—wealth of bracken
and clinging bramble—Stanmore Common is full of
interest: it is not a flat expanse by any means, but, seen
from the road, undulates in graceful lines down towards
Elstree, silver birch in the middle distance, and Scotch
firs crowning the heights beyond.

Bushey Mill that you see marked on the map must not
mislead you; it is a name given to a cluster of houses
standing on the site of a mill pulled down years ago, not
worth putting yourself out of the way to see; the same
remark applies to Otter's Pool, both "great" and "little"
as regards nomenclature: the names are misleading.

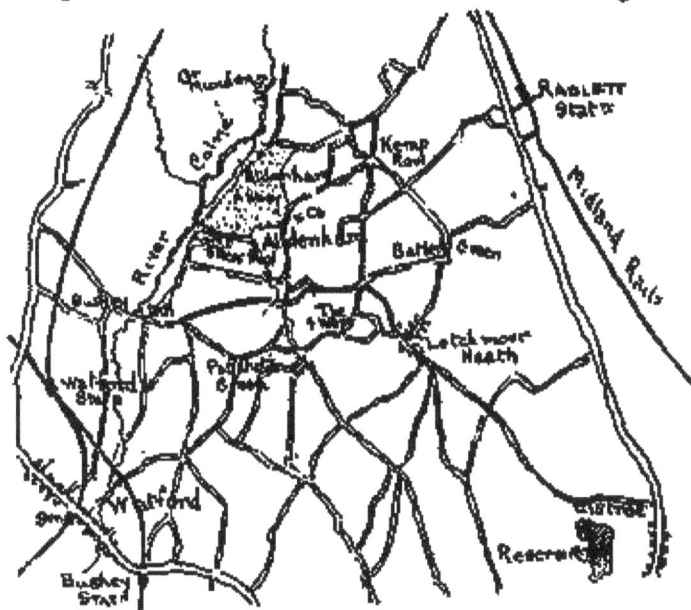

Scale, ½-inch to the mile.

You will have no difficulty in finding Aldenham from
either Bushey or Watford if you follow the map.   Aldenham Church is a very picturesque one, and, as regards
photography, fairly accessible before the foliage is fully
out.   With a long focus lens the best view is from a path
over the fields on the south-east side, as you are able to

get a long range of farm-buildings and hay-stacks to
break up the foreground, though there is another to be
had from the road, lined with beautiful elms, which is
almost as good. An afternoon light suits both these
views, and the last-mentioned should be taken not later
than April or May. Opposite the church is a very
comfortable inn called " The Chequers," and on the left
of it is a carriage drive leading to Aldenham Abbey ; the
public footpath which joins it later on is by a gate at the
right of "The Chequers." Take this path, and where
the avenue meets it, turn to the right across the field,
at the further end of which you see an open gate; the
manifold finger-posts will tell you which is the drive and
which the footpath, and as the two frequently come
together it will be well to follow their directions. About
half a mile from Aldenham Church you come to the
abbey. I find no mention of it in Cassell's "Greater
London," and it surprises me. It may be a modern
house—in fact it has that appearance, but it is certainly
a fine one and worthy of the photographer's attention,
with its castellated turrets and splendid situation. It
faces, as nearly as I can judge, due west, and a morning
light will not serve at all. The footpath passes right in
front of the lawn, and presently reaches the river. Here
is a finger-post at the junction of four ways : the one on
the right leads to Colney : that on the left (a bridle-path)
to Great Otter's Pool, straight on over the wooden bridge
crossing the river to Bricket Wood. You cannot take
the whole four ways at once, but they are all deserving of
attention in turn. Cross the water and trespass a hundred
yards or so to the right to see a little brick bridge

surrounded by trees—the reflections in the water perfect,
so still is this arm of the river; late in the afternoon it
makes a picture.

A little further, returning to the legitimate path, is the
river again, and another wooden bridge with something

A bridle path Aldenham Abbey.

in the way of a waterfall at the foot of it. After heavy
rains, or, at any rate, when there has not been a long
spell of dry weather, it has quite a respectable volume
of water, and under certain conditions is worth
photographing. Here the path forks, and standing on a
little hillock between the two is a yew tree overshadowing

a gravel pit; this is a favourite resting place for cattle,
and you can rarely see it without finding at the same time
four-footed friends of some sort. There was a flock of
sheep grouped under its shade the last time I saw it, and
although they went away on my setting up my tripod,
curiosity or good-nature got the better of them, and they
returned to see what I was doing. You can never be too
patient in photographing animals—Mr. Gambier Bolton
is a living exemplar of this patience—and it is wonderful
how soon their shyness is overcome and their fear of the
camera vanishes.

Let us leave this path, although it may be followed to
Bricket Wood with advantage, and return to the finger-
post across the river; one of the arms points to Great
Otters Pool, and through the first iron gate in that direction are some shaggy, long-haired Highland cattle. It
is wonderful how such little beasts can look so fierce and
be so harmless with it; they sniff the air at your approach
and watch your every movement as though undetermined
at present whether to treat you as a friend or an enemy.
Here again your patience may be exercised; the groups
break up at first as you get too near them, but by-and-bye
they ignore you, and with a shutter and a good light you
may work your own sweet will.

After having traversed the next field the path broadens,
skirting a spinney on the one side and the river with a
wooden bridge on the other; a delightful path it is, with
hedgerows that are wild and five-barred gates that would
satisfy an artist. You come across two or three houses,
the Great Otter's Pool before mentioned, then the path
becomes a lane winding into Watford, two miles further on.

There are so many different routes in this section of the map that I hesitate in recommending any special railway station to finish up with; for instance, having got so far as Aldenham, it will not do to miss Letchmoor Heath or Patchett's Green, they both have so many pretty little cottages, and the lanes connecting them with Aldenham and each other have much to recommend them to the photographer. Radlett is the nearest station for Letchmoor Heath, and there are pretty frequent trains to the city from there, although it is not such a good service as from Watford. Elstree is also on the Midland, and has an additional train or two—notably one at 5.55 p.m.

Elstree Reservoir is not of much account as far as I can make out, and you may spend a lot of time in going round it without finding a suitable view. Fishing is a different matter, as anglers will tell you.

To do this round comfortably and thoroughly a whole day is almost necessary; but if you attempt to do it on a half-holiday, let it be in midsummer and choose Radlett as your station, keeping in the vicinity of Aldenham Church and Aldenham Abbey. There is plenty of work to be done; in fact, with a limited number of plates, you will have to stay and consider what you can do without.

CHAPTER IX.

POTTER'S BAR, MIMMS, AND SHENLEY.

By Great Northern Railway to Potter's Bar.—
The way to North Mimms.—Mimms Hall and a
Winding Stream.—The Hamlet of Water End
and a Curious Cottage.—Through the Lodge
Gates to North Mimms Church and through
the Park to South Mimms.—Shenley and the
Road to Radlett. — Shenley Church and
Colney Chapel.—A Water-Splash and a Happy
Composition.

THE nearest railway station to North and South
Mimms is Potter's Bar, reached by the Great
Northern from King's Cross or the lines connecting with
it at Finsbury Park. It is rather a long way out, perhaps,
but not too far for a Saturday afternoon's excursion on a
summer day, and may very well be included in this series
of photographic rambles.

At Potter's Bar Station, instead of turning under the
bridge in the direction of the town, take the other way
as far as the first finger-post, only a few yards up the road.
It directs on the right to North and South Mimms *via*
Mutton Lane, a good road at most times, and very
charming in summer. About a mile and a half up it
takes a sudden turn in a northerly direction, then at the
foot of a slight descent you come across Mimms Hall,

now a farmhouse. This is worth looking at for its gables
and haystacks surrounded by tall trees, mostly poplars,
though I cannot say it has sufficient in it to justify the
exposure of many plates; the photographer will judge
for himself. You will notice, winding side by side with

Scale ⅜ of an inch to the mile

this pretty lane, a little stream; it is a branch of the
river Colne, and in many places is very picturesque,
whether swollen by heavy rains or nearly dry from long-
continued drought. The road at the left a little further
on leads to South Mimms, and you get a bird's-eye view
of the village with its solid-looking church from the

junction of the roads. Leave this for the present, and
keep straight on to Water End, an outlying hamlet of
North Mimms, where there are some wooden bridges
across the stream and an old gabled cottage on the left
just before entering the village. It has recently been
painted and otherwise "done up," so that much of its
quaintness has disappeared, but the last time I saw it,
snow lay thick on the ground, and the house with its
white roof against a grey background of trees made a
very effective picture, heightened by the presence of an
old lady—presumably its occupant—in the doorway.

Make your way through the village, a very small one,
and you will see on the left, a little beyond the schools,
a path over the fields to the church. There is another
way, a little further on, too, through the lodge gates, and
perhaps this is better, because from the rising ground
you get such a charming glimpse of it. The buttressed
tower and slender spire, and the ivy-covered walls,
" arrange " very well through the trees, but it is best to
secure this view before the leaves are fully out, or much
of the outline is lost. A morning light, say in April or
May—time 10 or 11—is most suitable from this point,
and an afternoon for the other side of it. It is a delightful
little church built of flint and stone, the roofs well broken
up and characterised as a whole by very graceful proportions. The sketch gives a good general idea, but it
is not taken from the point I have described, as it should
have been.

The churchyard is very pretty, too; a splendid elm
at the further gate is followed by an avenue of limes up
to the hall ; but before reaching it the public thorough-

fare turns round to the left through one of the gates of
North Mimms Park, passes a water tower, and then over
hill and dale and undulating expanses of bracken it runs
through woods to South Mimms about two miles further
on. Never mind the gates, except as regards shutting them;

North Mimms C⁴

neither let the notice boards referring to trespassers
disconcert you—they only refer to the woods and other
people than photographers; the pathway is public.

South Mimms lies at the bottom of the "old road"
after emerging from the park, and you get a glimpse of
it several times before leaving the woods. It is a curious
little village, and in some respects a pretty one. The

church is substantial-looking, with tower and turret and
clinging masses of ivy; look at it from the road at this
side, facing apparently north-west: a four o'clock summer
sun lights up its massive proportions, and if you can get
far enough back, or as far as the road and your lens will
allow, there is a picture to be made.

From South Mimms no railway station is nearer than
Potter's Bar, and the road to Barnet is not very interest-
ing; but the one to Shenley (which you will notice
marked in the bottom corner of the map) is very pretty.
There is a more convenient station for it, however, than
Potter's Bar—this is Radlett on the Midland line.
Coming from here the road cuts through the estates of
Porters and New Organ Hall—two curious names enough
—and is very pleasantly shaded by trees all the way to
Shenley.

I have seen Shenley under a recent and severe attack
of hoar-frost: the hedges like frosted silver, every roadside pond a study, and every little clump of wayside
rushes thrusting their sharp spikes through the glistening
snow—with just that faint mistiness in the atmosphere
that lends "distance," makes a photograph a picture,
and sets one puzzling to find out whether, after all,
Shenley is not better worth a visit at this season than
any other. The church—the old church, not the chapel-
of-ease in the centre of the village—is situated fully a
mile to the north in the direction of St. Albans, and is
really at Shenley Bury. There is a pretty lych-gate to
it, and a cluster of cottages down the road, with a farm-
yard, make a pretty background, but I do not think the
photographer will think the church worth taking. The

old illustrations of it, showing a white wooden tower at the south side and a solid-looking yew tree, are misleading; the tower has been pulled down for some years, and the yew tree assumes more modest proportions on closer acquaintance.

If it would not make the round too serious to be undertaken when burdened with a camera, it is better worth while to go on to Colney (the map breaks off at this point, but the road continues until it joins the main road a mile or so south of St. Albans); this will bring you to Colney Chapel, where a branch of the river Colne, lending its name to the various Colneys hereabouts, crosses the road and forms what is known as a "water-splash." A wooden footbridge serves for passenger traffic, and the view it presents is far from unpleasing. Still, it wants something to complete it as a picture, and this "something" is worth waiting for. I took this into consideration one sunny autumn day, having fixed my camera in a commanding position—occupying myself that of Mr. Micawber. An opportunity was not long in presenting itself, for coming down the lane from St. Albans was a lumbering hay-cart, lazily drawn by a team of three, languidly driven by one—if a sleeping son of the soil may be said to drive—completing an arrangement more satisfactory than expected.

If you are unable to follow this lane right away to St. Albans, three miles ahead, it will at any rate repay you to traverse it for a little distance to see what a charming lane it is, and how refreshing its broad grassy sides are to the eye of a photographer.

CHAPTER X.

PONDER'S END, WALTHAM, ENFIELD LOCK.

On the Great Eastern Railway.—The River Lea: a digression on marshy land.—Ponder's End. —Enfield Lock. — Waltham. — The Abbey Church: view from the river.—An old Gateway.—The Manor Farm and Harold's Bridge. —The Interior of the Abbey.—The Cattle Market: a quaint bit of timbered architecture.—Various Branches of the Lea.—Back from Enfield Lock.

THE course of our photographic rambles has brought us to the Great Eastern Railway, a line particularly suited for their continuance on account of the lovely country it passes through, as well as for its admirable train service and cheap return fares. Let us take the main line first; from Liverpool Street to Clapton it is devoid of interest, but the latter station passed, we have the first glimpse of the River Lea, so loved of Izaak Walton and the followers of his gentle craft, and the marshes through which for miles and miles it takes its sinuous course. To digress for a moment : there seems to be amongst some amateurs a lack of appreciation of marshy land and low-lying country, but with careful treatment, and the exercise of judgment, it should yield many good pictures of skies and wind-swept commons;

where a river runs, especially, there should be no end of possibilities in the way of picturesque combination.

All the country in this neighbourhood is flat, Ponder's End, Enfield Lock, reminding one forcibly of Dutch scenery, with solitary poplars and groups of other

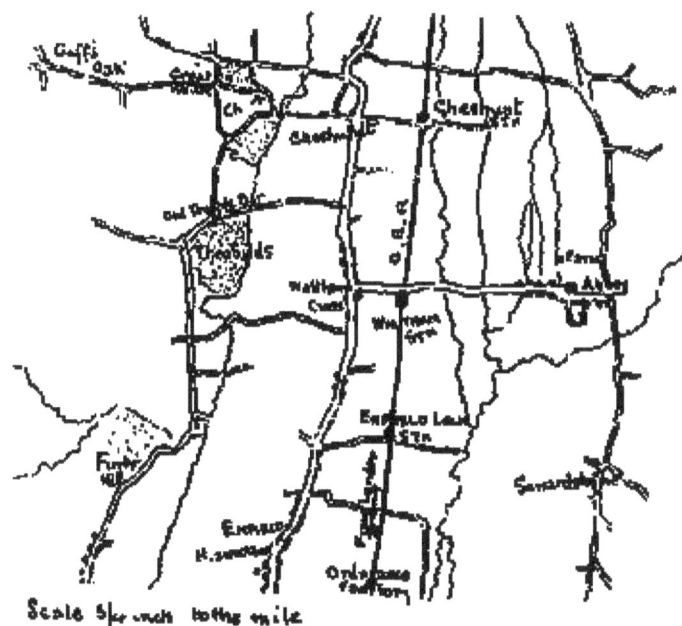

Scale 5/8 inch to the mile

trees in twos and threes. These places are desolate looking enough, but for photographic work of a special kind they are distinctly suitable. To explain better what I mean, look what may be done and has been done with the Broads of Norfolk, the marsh lands of Essex, and the dreary Canvey, expanses as flat as any in

E

England; see what pictures may be made of desolation, combined with a judicious use of sky negatives.

The digression has been longer than I intended, but the temptation to say something on behalf of a flat country overcame me. Ordnance Factory comes next, and the name explains itself; then we reach Waltham, thirteen miles out.

Coming from Waltham Station you cross the railway, *i.e.*, instead of taking the left hand towards Waltham Cross, you turn to the right in the direction of Waltham Abbey. It is a mile into the town, by a dull, uninteresting road lined with houses on each side ; you cross the canal and one branch of the River Lea and find yourself in the midst of one of the quaintest towns it is possible to conceive, full of gabled houses and timbered gateways, brick, wood and stone in heterogeneous masses everywhere. The first view you get of the Abbey Church—near view I mean—is from this road; the foreground consisting of the houses before-mentioned is pleasing, and the whole comfortably to be got in on the long way of the plate. Taken altogether, I do not know whether the church may be considered strictly handsome; it has been so restored and re-built, and so much of its former grandeur has departed, that there is a hard, stern, set look upon it which the photographer would alter if he could. There are points from which a charming view is to be had of it, but it is with the abbey as an accessory rather than the chief point of interest—such a view is that from the banks of the river on the north side, taking in the fine old gateway and bridge, which forms the subject of our sketch. The gateway is partly stone, partly brick, and the road

through it skirts the remains of the old abbey walls, past
a long range of brick buildings which must have been
at one time the old manor farm, until it reaches the
road beyond, connecting Waltham with Nasing. I
would strongly recommend you to follow this road, as
it is full of interest, and you can by turning to the
right regain the town, making a complete circuit of
the abbey.

The exterior of the church seems to be in various
styles, in good condition and otherwise; its tower,
standing at the western end, is very substantial and
partially covered with ivy. Massiveness seems to be the
chief characteristic of the church both inside and out;
the columns are very heavy, some in spiral, others in zig-
zag form; the windows are small, and are filled with
stained glass, giving a general idea of gloom and op-
pressiveness to the whole structure. Looking up at the
ceiling from the nave gives you the impression that it is
tiled, but, when you ascend the winding staircase at the
left of the main entrance and reach the gallery,
you find that it is painted. On enquiry you will be
told that it was the work of E. J. Poynter, R.A., and
is a duplicate of the one at Peterborough. Very little
of the crypt remains—in fact, only that portion which
lies under the Lady Chapel, which has been recently
restored: what there is left serves partly for a coal
cellar.

I should have mentioned that the keys may be had
from Mrs. Knight, the caretaker, whose house is a few
yards down Church Street on the right hand, and next to
a bootmaker's: a charge of sixpence, devoted to church

expenses, is made for inspection. On Saturday afternoons the door is generally open for cleaning purposes, and though that fact will not let your off you sixpence, it will save you the trouble of fetching the keys.

On no account miss the cattle market, which lies off at the right, after leaving the church. I know of no

Waltham Abbey.

quainter bit of gabled timber architecture near London, unless it be the old horse fair at Kingston, which, by the way, is shortly coming down. Photographs of such places will be valuable in years to come. Through the cattle market you reach the river, or one branch of it, and its banks are worth following for some distance, for instance, there is an old farmhouse (the manor farm

before mentioned) with a one-arched stone bridge,
crumbling and decaying, that connects the farmyard with
the meadows on the other side of the river ; it is known
as Harold's Bridge. Whether it had anything to do with
Harold or not is a matter of conjecture, but it is very

Harold's Bridge (Waltham

picturesque with its pollarded willow and the roof of the
farm buildings behind.

The various branches of the Lea—traditionally supposed to be the work of Alfred the Great in his endeavour
to leave the Danish fleet aground by cutting new channels
—are chiefly open to the public, and an hour or two might
be profitably spent in exploring them ; they merge below
Waltham into the Lea and Stort navigation and traverse

some little distance before they are separated again.   It
is almost worth while to follow the banks of the canal to
Enfield Lock, the next station nearer London, passing on
the way the Government powder mills on the further side ;
there are many quiet back-waters and locks, and the fact
of the Lea being an angler's river, *par excellence*, furnishes
you with figure studies whose patience in sitting is
proverbial.

WALTHAM CROSS, CHESHUNT.

CHAPTER XI.

LIVERPOOL STREET TO WALTHAM.—ELEANOR'S CROSS.—
THE "FALCON" AND "YE FOURE SWANNES."—
CHESHUNT.—THE BURIAL-PLACE OF THE CROMWELL
FAMILY.—THE "GREAT HOUSE."—GOFF'S OAK : A
RUINED TREE AND A WINDMILL.—"THEOBALD'S,"
ANCIENT AND MODERN.— TEMPLE BAR : AN OLD
FRIEND WITH A NEW FACE.—THE NEW RIVER, AND
BACK TO WALTHAM CROSS.

WE make for Waltham Station again by the route mentioned in the last chapter; but instead of crossing the railway to Waltham Abbey as before, we turn to the left and strike out for Waltham Cross. It is only a quarter of a mile away, and is worth a visit for many reasons : First, it is one of the only three crosses remaining out of the ten planted by Edward I. in memory of Queen Eleanor on those spots where her body rested on its way from Hareby in Lincolnshire to its last home in Westminster Abbey. This is history, but the landlord of "Ye Foure Swannes" is prepared to knock the legend on the head in a most summary manner—by documentary evidence if necessary. However, that does not matter to the photographer. I do not propose to detail the history of this cross—in fact, I am not sure that I know it sufficiently well to give reliable information. Mr. Tydeman, of the "Foure Swannes," is an antiquary of the first order, and took an important part in the restoration of the cross some years ago. It must then have been in a very dilapidated condition, but the labour that was devoted to it has made it presentable again. You will see that much of the old has been replaced by new with as much accuracy, historical and architectural,

The Falcon, Waltham Cross

as the existing records afforded. As you come from the station the cross looks insignificant. This is because the hostelry of the "Falcon" has been built too close up to it to give it breathing space. The inn on the other side of the road lays claim to greater antiquity, and the sign-board of the "Foure Swannes," stretched right across the road, is very striking. The board is modern, of course, but it has been modelled on the old design, which it has replaced piece by piece. The spelling has been preserved as well, and one is ready to forgive an apparent imposition on account of its picturesque aspect.

It is not very far from Waltham Cross to Theobalds, about which something will be said presently, but it is almost better to take train to Cheshunt (the next station further on), and reach it from there. The whole neighbourhood is full of pretty little bits, and the village of picturesque cottages; the church is a very good specimen of flints and stone, though suffering (from a photographer's point of view) from recent restoration, and the graveyard contains some interesting tombstones, notably that of the Cromwell family), in the north-west corner. You can get a photograph of the church from almost all quarters, and, seen from a distance through the many trees that surround it, it makes a very pretty picture. Through the churchyard, following a gravel path over a field or two—notice the quiet pond with its fallen tree near the church—you reach the road from Cheshunt to Goff's Oak and Cuffley; turn to the left and you will see on a little eminence, across the road, a solid brick building, known as Cheshunt Great House; it is supposed to be the old manor house, but at first sight you are wondering whether it is a chapel or a barracks, as bare and ugly a building as it is possible to conceive. The interior is more interesting, with its old pictures and tapestry, and may be seen on application any day except Sunday; the entrance is round by the back, a part of the house in occupation by the caretaker and his family. A mile and

a-half west of the Great House lies Goff's Oak, a part of
Cheshunt Common, taking its name from a battered and
decayed specimen of the tree standing opposite a public-
house at the cross-roads ; its cavities are boarded up to
keep out wind and weather, and its immensity has been
very much exaggerated, so that I do not think the photographer will care to expose a plate upon it.

   I have mentioned Goff's Oak because it boasts a
windmill, one of the very few I know on the north side
of the river ; there is another at Barnet Gate, between
Elstree and Barnet, and one also at Enfield.   The country
side all round here is very pretty, full of bridle-paths and
quiet lanes, which should be explored diligently.   You
will see on the map, nearer Waltham than Cheshunt, a

park called "Theobalds"; it was once a royal residence, but before that it was the family seat of the Cecils. James I. took a fancy to it, and gave Hatfield to the first Earl of Salisbury in exchange. Like many more of the great historical mansions, hardly a stone of the original remains, and the site is occupied by a less pretentious house in the possession of Sir Henry Meux; there is a public footpath through the park from which you obtain a glimpse of it as it now stands—nothing in comparison with the old palace, if pictorial representation is to be believed.

What makes Theobalds especially worth the photographer's attention, is the presence of old Temple Bar rusticating at its northern entrance. When Fleet Street was disencumbered of the grand old gateway, each stone was numbered before it was taken down, but there is a dreadful legend to the effect that the numbering was done in water colour, and that as the mass lay in Battersea Park, prior to removal, a heavy shower came down and washed the numbers out. What does it matter, after all, if the story should be true? There were sufficient photographs and drawings in existence to make the task of re-erection easy enough, and old Temple Bar stands in all its integrity, more imposing than ever, in its quiet suburban retreat—an old friend, with not a new face exactly, but with different surroundings. The New River runs through Cheshunt and skirts Theobalds. Much of it is open to the public, and, though prim and straight-cut in some places, is sufficiently pretty to tempt one to stray away from the roads. The nearest station from here is Waltham Cross.

CHAPTER XII.

EPPING FOREST

CHARACTERISTICS OF THE FOREST.—CHINGFORD AND
THE WAY TO REACH IT.—QUEEN ELIZABETH'S
HUNTING LODGE AND THE "ROYAL FOREST
HOTEL."—CONNAUGHT WATER.—THROUGH MODERN
CHINGFORD TO THE OLD CHURCH.—A PICTURESQUE
COTTAGE AND WAYSIDE POND.—A FINE RUIN.

IN the chapters under this heading I shall endeavour to
give a general idea of the value of each particular
district from a photographic standpoint, omitting ancient
history as much as possible. There are so many standard
works of reference on Epping Forest that I hesitate to
recommend any special one for study. The story of its
preservation, and the description of its flora and fauna
are interesting reading, and the photographer should have
some knowledge of them, if only superficial.

The whole of the Forest is free to the public, and the
Great Eastern Railway Company have encircled it at
almost every point permitted them to touch. There is
an excellent service of trains, and cheap return tickets
are issued from early spring to late autumn to Chingford
and Loughton, available on the home journey from either
station. Good roads are cut across it in every direction,
and the numberless paths that wind through tangled

HIGH BEECH WOODS, ESSEX.

bracken and thick-set briars, under sloes and crab-apples, hawthorn and honeysuckle, afford the photographer and the naturalist many a pleasant and profitable ramble.

The distinctive feature of the older forest—its continuity—has disappeared, through encroachment from time immemorial, and although some fine trees are yet to

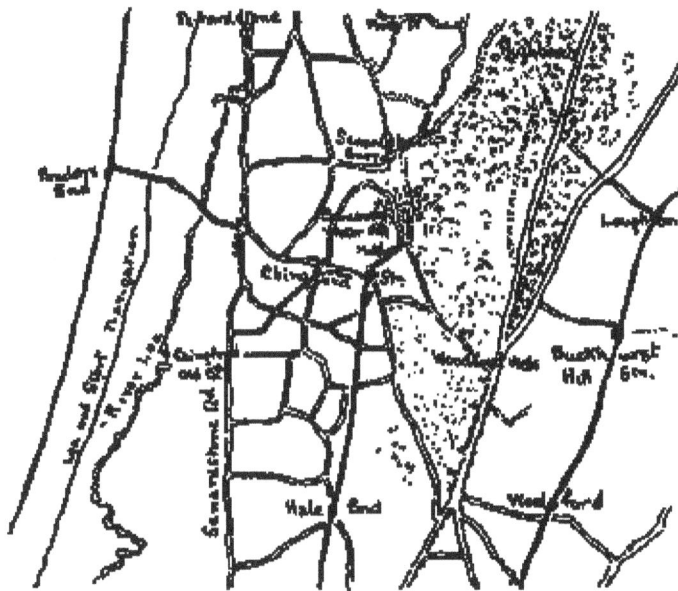

be found—beeches at High Beech, grand old oaks at Chingford and Bush Wood (a part of the lower forest near Leytonstone)—the general character of the woodland is somewhat disappointing on account of its stunted growth. Hornbeams, which are scarce in some parts of the country, are here plentiful, but the timber of Epping

Forest consists chiefly of pollarded beeches. Deer are to
be found occasionally. They are supposed to number
between 130 and 140, but are thinned out at certain
seasons; and, as far as my own experience goes, they are
not unusually shy.

The two best-known parts of the Forest are un-

Chingford Old Church

doubtedly Chingford and High Beech, but it more than
repays the photographer to keep in the less frequented
localities such as Theydon Bois, Lippit's Hill, and
Mott Street, Sewardstone and the Wake Valley, in
order to do good work, and to do it in peace. Let
us begin with Chingford, the usual starting point,
and almost the least interesting. Close to the station

is the "Royal Forest Hotel," the shrine of many
a forest pilgrimage, standing on an eminence that rises
abruptly from Chingford Plain, a handsome structure
enough, built to conform with the architectural style of
Queen Elizabeth's Hunting Lodge that is dwarfed by its
too great proximity. This latter is of wood and brick,
with gabled ends and a high-pitched roof in true Elizabethan style, and, though in a good state of preservation,
lays claim to undoubted antiquity; its chief interest is to
be found in the interior, in the oak dining-room, massively
timbered and hung with timeworn tapestries, as well as
the broad staircase leading to it, up which Queen
Elizabeth is popularly supposed to have ridden on horse-
back. In the glades opposite the lodge and hotel are
some really fine tree specimens; "the gnarled oak flings
his mighty limbs on high," and in his shade the itinerant
photographer does a thriving business, while in the brief
intervals permitted them, the much-abused beasts of
burden, ponies and donkeys, seek that repose which the
exigencies of the public service so rarely afford.

Close to the hotel, also, is the Forest pool, better
known in late years as Connaught Water; it has an
artificial appearance, with its islands and many pleasure
boats—including a steamer, unfortunately—but the background of the Forest is sufficiently charming to warrant
the exposure of a plate.

Chingford is a large parish, and the newer village,
with its modern and conventional church, may be dis-
missed without special mention; not so the old church,
which lies a good mile and a half south of the railway
station, adjoining the cemetery. Hale End Station seems

almost as near for it; but coming from Chingford you pass the new church, and a few yards further on a finger-post directs, by the road turning abruptly to the left, to Walthamstow and Edmonton; this is the road you must follow. Notice about a quarter of a mile further on a picturesque cottage at your left, and then, as the road winds to the right again, you will see a very dilapidated barn with a pond in front of it, the resort of many and various ducks. You cannot mistake it, for it stands at the junction of the lane leading to the ruins of the Old Church, whose outlines, from Birket Foster's fancy sketch of many years ago to the latest representation in the form of a Christmas card, are so familiar to Londoners that I am almost afraid to give a sketch in these columns. The photograph, to do it justice, has, I think, still to be produced; first, it is rather difficult to approach, in spite of its standing on the crest of a hill: secondly, it is so covered with masses of ivy, that "halation" usually ensues unless special precautions are taken to avoid it; there is, however, no just reason why the view should be taken from the roadway at the south-east side, as it almost invariably is. Such a delightful picture should be studied carefully; brick, stone, flint, and plaster are so jumbled together, and the local colouring is so rich, that it cannot fail to please the eye of the most fastidious. Look at the brick porch at the south side, with two immense roots of ivy clinging round it, hugging it so closely as to swallow up almost all the architectural detail, except the doorway, in its upward growth; how, having obtained a firm footing, it clambers up to the roof of the porch, and steals from there to that of the church, filling up every niche in its grasping embrace until it reaches the tower and the weather-beaten mast that surmounts it. All the glass is gone, and the stone-work is in the last stage of decay; the plaster is falling from the walls, showing the red brick beneath; you can look through the crumbling windows and see that the interior is bare of everything, and the arches are shored up to prevent premature collapse. The graveyard, with its grass-grown paths and lichen-covered stones is a study in itself, and quite in keeping with the old church. The outward shell of Chingford Church will not stand the stress of weather many years longer, but its memory will live for generations.

Angel Road Station on the Waltham line is almost as near as Chingford, and you can see—looking across the valley of the Lea—the tall chimneys of Edmonton and Tottenham, with the smoke of London in the distance; but, before leaving this locality, turn down by the foot-path over the fields and reach the river. On its banks are many studies—a wooden bridge or two and cattle innumerable, to say nothing of that low-lying marshy land and reedy water that are of such value in composition.

CHAPTER XIII.

EPPING FOREST.—*Continued.*

SEWARDSTONE AND ITS CAPABILITIES FOR PICTURE
  MAKING. — SEWARDSTONE MILL, AND RIVER
  SCENERY.—THE VALLEY OF THE LEA: WOODEN
  BRIDGES AND MARSHY LAND.—PLOUGHING.—THE
  ROAD TO WALTHAM ABBEY.—A TUMBLEDOWN
  BARN.—TO LIPPIT'S HILL AND HIGH BEECH.—
  "THE OWL."—A VIEW OVER THE FOREST.

A CORRESPONDENT wrote to me on the publication
of the Waltham Abbey chapter to ask why I had
done such scant justice to Sewardstone. As a matter of
fact I had not done it at all, but was intending to include
it in the Chingford district; my friend, however, has
described it better than I could have done myself, so I
will quote from his letter:—" The marsh around Sewardstone contains some of the finest things in the county;
there are some good hill crests on the Sewardstone side
which make satisfactory backgrounds—tortuous, reedy
rushy river, with some three or four delightful wooden
single-railed bridges, to say nothing of the clusters of
willow trees, the cool green of which is so finely broken
up with red-roofed cottages. Then about the cattle,
sheep and horses: I know a particularly fine herd of
cattle which at 4 p.m. every day in Spring and Summer
a quite picturesque old dame and children come and

drive home over a ricketty bridge, and a fine picture is made when the afternoon sun is getting low. That's just close to the 'Fox and Hounds' on the Sewardstone and Waltham Road." This is so; and if my correspondent will allow me to supplement his valuable contribution, I would add that just beyond the "Fox and Hounds"—say two or three hundred yards—there is a gate and a footpath through it leading to a group of cottages and Sewardstone Mill, and that within a space of a quarter of a mile or so there is sufficient work to keep one employed for an afternoon—and a long afternoon, too. Leaving the cottages on the right, the footpath cuts across diagonally to the river bank and then crosses a wooden bridge that spans one branch of the Lea coming from the direction of the mill; but the chief charm is found in the many views that the river itself presents. In some places it is dry, and leaves the gravel bed exposed and clumps of rushes aground while it seeks fresh channels burrowing under the overhanging banks; if one could only get some lazy cattle here on a summer afternoon any number of pictures could be obtained with their assistance. A mill wheel is standing a few yards away, but as there is not a vestige of any other building to back it up, its appearance is puzzling to say the least of it; it seems to be in fairly good condition, too.

But let us retrospect a little. In the last chapter we finished up at Chingford Old Church, and the footpath that leads from it to the valley of the Lea. I want you to follow this path over a couple of meadows—not only for the sake of the view you get of the church almost

surrounded by its tall elms, but in order to reach the
river. There is a well-marked right of way for a long
distance, trodden by the feet of many generations of
anglers, whose river it is, and you follow the path in the
direction of Sewardstone; the water, sedge-grown and
shaded by many willow trees, is as placid and peaceful

On the Lea near Chingford.

as is the occupation of the figures on its banks. Have
you seen Dendy Sadler's "A pegged-down fishing
match"? I know all those figures, the man who is
throwing in the "ground-bait" amongst them—and they
are all Lea anglers—possibly Rye House and Broxbourne
men, unless they come from further afield—the Waveney
or the Bure.

About half a mile on we reach a wooden bridge,
which is worth looking at from the other side. I have

given a rough sketch of it, because the landscape strikes
me as being very pleasing : the reedy river with its long
floating masses brought to a standstill against the wooden
piles ; a plank crossing the dyke that feeds it from the
Chingford side, with a fallen tree beyond it—it only
wants that old dame and her cows afore-mentioned to
break up the lines of the bridge, and we have a picture.

All the marshy land round here is fairly open, and.
except you are brought up by a dyke that heavy rains
have rendered impassable, you may wander where you
will. On the other side of the bridge (*i.e.*, the Chingford
and Sewardstone side), a footpath bears to the left across
the meadows and then over a ploughed field, where at
this season of the year you may generally find a plough-
man and his plodding team—I rather think they are
harrowing just now, but that is a matter of detail. It is
a spot where Mr. Gale's advice to make yourself agreeable
to the man at the plough, watching at the same time for
the particular spot in which he and his team best
compose with the landscape, comes in useful; never
being in a hurry with the day before you. At the other
end of the field the path cuts off to the right to join the
road from Chingford through Sewardstone to Waltham
Abbey, a road that runs almost parallel with the River
Lea almost the whole way ; along it you will find many
pretty bits, fine elms and roadside ponds, with here and
there a farm yard and cattle at the gate waiting for
admission (usually about 4 o'clock), until you reach the
" Fox and Hounds." Then comes the footpath I
mentioned in the early part of the chapter, to Sewardstone Mill, while the road turns abruptly to the north ; a

little further on and a road branches off at the right through Sewardstone Green and Sewardstone Bury, which you can follow to Chingford Station if you are inclined to return so soon. Keeping straight on—it is barely a mile—you will reach the "Royal Oak" and a cluster of cottages; beyond the inn, a few hundred yards, there is a barn that will delight you: it stands in a farmyard, naturally, but it is emphatically the worse for wear. One end is open to all the winds of Heaven; the sides are of wood, green with age, rotten with weather; while the red tiles, fluted and plain, are in graceful curves and well broken-up lines. A heap of straw and refuse on which every cock in the yard crows in turn, until dispossessed by a later comer, stands in front of it; a broken-down cart or two and some cattle. Well! you have a picture there better than I can describe.

A finger-post directs on the right past this ruined barn to Lippit's Hill and High Beech; it is a long and steady climb of a mile and a half to "The Owl" at the top, but there is such a charm in this lane skirting the Forest, that it makes one disposed to linger over the bye lanes and bridle paths that branch from it. A quarter of a mile from the top you can get a breathing space, by looking at a very picturesque cottage standing with some out-buildings, a five-barred gate and a haystack or two, on the right. Notice the way in which the tiles run in a marked concave, and how from a certain point the whole composes very favourably.

On the crest of the hill is a little pond, then the red-tiled roofs and timber walls of "The Owl" come in sight. From this point the extensive panorama of the

Forest is very striking ; Buckhurst Hill and Loughton across the valley, High Beech on the left, Chingford at the right, and down at your feet Fairmead Plain—forest on all sides of you. The station at Chingford is about a mile and a half away, and may be reached by the road that runs at right angles to the one coming down from "The Owl" by turning to the right, but it is no further and more agreeable to traverse the Forest by way of Connaught Water and the "Royal Forest Hotel" to arrive at the same destination. A map is very useful in Epping Forest, and an excellent one on a large scale at a small price is published by Bacon & Co. ; it costs threepence, and may be obtained at most of the railway bookstalls on the line.

CHAPTER XIV.

EPPING FOREST.—*Continued.*

HIGH BEECH CHURCH.—THE QUEEN'S OAK.—A VIEW
ACROSS THE FOREST.—THE WAKE ARMS.—AMBRES-
BURY BANKS SEEN PHOTOGRAPHICALLY.—EPPING:
THE OLD CHURCH TWO MILES NORTH OF THE
TOWN — THE TOWER A GRAND STUDY.— EPPING
UPLAND AND NAZING WOOD COMMON : A PIC-
TURESQUE CORNER OF ESSEX. — THEYDON BOIS
AND A FOREST ROAD.—CATTLE STUDIES.—A SCENE
IN THE WAKE VALLEY.

THE last chapter brought us within sight of High
Beech ; the sharp descent from "The Owl" on
Lippit's Hill led down to Church-road and Whitehouse
Plain, while Fairmead Bottom lay in the valley before
us, and the church in the forest crowned High Beech.
Making our way up the next hill we see that it is a
modern church, built of white stone, and though hardly
picturesque enough to make a photograph by itself ; seen
through the forest glades, with a foreground of wild briar
and bracken, it forms by no means an unpleasing accessory. The shape of the spire is peculiar, and built of
the same material as the body of the church. Its commanding position makes it serve as a landmark for the
whole county. Hard by is the Queen's Oak, planted by Her
Majesty to replace the Old King's Oak — long since a
a victim to stress of weather—on the 6th of May, 1882,

when the Forest was formally given to her good people
for ever. It was practically public property before, but
this event put an official stamp upon the freedom of
the Forest never to be effaced.

Behind High Beech Church a gravel drive cuts
through a more open part of the Forest, and from it

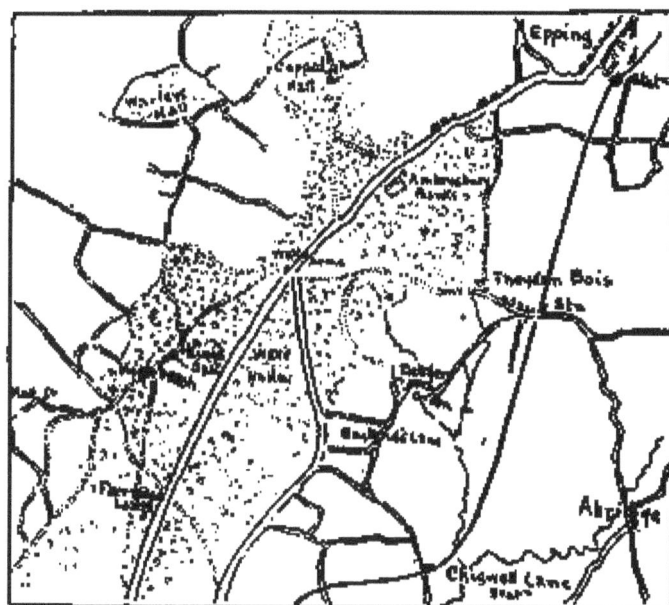

Scale No inch to a mile

you obtain a splendid view over the valley of the Lea,
with Waltham Abbey in the distance. Two miles north
is the "Wake Arms," a well-known hostelry, and here
the new road through the Forest joins the main high-
way from Loughton, and merges into it. Opposite is

the road to Theydon Bois, on which I shall have something to say presently; but in the interval let us look at Epping, still another two miles further on. Before reaching it you will find about a mile and a half south of the town the remains of an old earthwork, known as Ambresbury Banks; as to its being Roman or British the antiquaries are not agreed, but so far as the photographer is concerned, it may be said that what there is to be seen is simply a raised wall of earth of irregular construction, grass grown and surmounted by slender trees, whose shadows reflected in the water that fills the fosse, help to make it a picture. It is nothing out of the common, however, and any of the little Forest pools between here and Chingford on the one hand, and Loughton on the other, are more effective studies. Epping is a long straggling town, nearly a mile long, with a station of its own on the eastern outskirts; a town of many inns, but of no great beauty apparent at first sight, it hardly appeals to the photographer's close and lively attention. The houses are modernised, and the inns re-built, but the town must have been of some importance in the old coaching days, situated as it was on the road to Newmarket and the eastern counties, yet infested with highwaymen and footpads, who made its name a by-word amongst travellers, timid or otherwise.

The water-tower in the centre of the main road might very well be mistaken for that of a church except at close quarters. Epping Church—the old one, not the modern, which is still unfinished—lies two miles or so north of the town on the way to Parndon and Nazing, in the heart of a purely agricultural country abounding

in picturesque and gabled farmhouses; in fact, I know
of no neighbourhood that would better repay the photographer for a week's sojourn than this. The absence
of a railway may be a disadvantage in some ways, but it
has left a great tract of country untouched and its natives
unsophisticated. We can only take a rapid survey of
it here, as it lies outside the bounds of greater London,
but there is sufficient material for many days to be
found in Epping Upland, Nazing Wood and Thorndon
Commons, Epping Bury and Parndon, great stretches
of picturesque Essex that one ought to know thoroughly
and photograph systematically.

Epping Church is a fine old building from a photographic stand-point ; the nave is, perhaps, out of
proportion to the massive tower, and the walls are
plastered with stucco, but the red brick ivy-covered
tower is magnificent—a study of itself. Some of the
bricks are missing, but the abundance of ivy, the
windows, and the massive yet graceful shape of the
whole structure make it worth more than cursory inspection. The churchyard, too, is very open and large, so
that there is plenty of room to move about in the
selection of the best view ; as to lighting, the tower faces
west.

Theydon Bois is the station before Epping on the
London side, and it saves something in time to take it as
a starting point to walk to Ambresbury Banks, or to the
"Wake Arms" ; it is a pretty little village with a
triangular green lined with trees. Notice the rustic farm
with its gabled barns and outhouses on the right as you
come from the station. Just before reaching the church

the road forks, the right hand going to Epping, the
finger-post on the left directing to Waltham Abbey. This
last is a delightful road; it cuts across a part of the
forest very little traversed by Bank-holiday crowds or
trippers of any kind, and the photographer has an
opportunity of working uninterruptedly. It dips down

At High Beech

into the valleys and over the rising ground with forest on
each side of it, numberless cattle, cows with their bells,
and shaggy ponies standing knee-deep in wealth of
bracken and forest grass that cover acres and acres of
these more open stretches of woodland scenery. This is

the place for cattle studies as pictures, because the surroundings are so beautiful and lend themselves to
successful composition ; you have fern and furze in the
foreground, and in the distance you have the blue haze of
the forest broken here and there by the delicate un-named

At High Beech

green of the lichen-covered oaks and the gleaming white
of the silver birch. You are not by any means confined
to the road either ; the woods at the left are some of the
finest in the Forest, broken up by miniature valleys and
beech-crowned heights, intersected by little-trodden
paths, and watered by numerous tiny pools. This is an

altogether different Epping Forest from the one we know in the neighbourhood of busy Chingford.   One picture— nay, there are many—especially appeals to the photographer in the Wake Valley; a little rivulet runs through it, its margins rush-grown and studded with pollarded beeches, on the crowns of which the *polypodium* is to be found; the silver birch overhangs it, and its slender form is reflected in the quiet pools, the image broken up by sharp spiked sedge and rippling shallows.

CHAPTER XV.

THE LOWER FOREST.

WANSTEAD PARK AND HOW TO GET THERE.—PER-
MISSION FROM GUILDHALL NECESSARY FOR PHOTO-
GRAPHING.—BUSH WOOD AND ITS AVENUES.—THE
ENTRANCE TO THE PARK.—ITS VARIOUS PONDS.
ROUND THE GREAT LAKE.—THE RUINED GROTTO,
AND A CURIOUS TREE.—THE HERONRY AND ITS
REED BEDS.—A VIEW ON THE RIVER RODING.

WANSTEAD PARK is not so well known as it deserves
to be, and this is all the more lamentable because
it is really such a beautiful place that one finds a want of
adjectives in describing it.

Permission to photograph is necessary, but it is easily
obtained by applying to the Town Clerk, Guildhall, E.C.,
for the requisite authority. By the same post this
courteous official will inform the head-keeper that your
name has been put upon the list of favoured individuals
to whom "permits" are accorded, and, having identified
yourself with your letter of credit, no further difficulty
will be met with.

There are several ways of reaching Wanstead Park.
I think about the best is by train from Liverpool Street
or Fenchurch Street to Leytonstone (the North London
has a connection between Victoria Park and Stratford),

and from there making your way across Bush Wood, a
part of the lower forest in which are two very fine
avenues, the last of those which we are told radiated in
every direction from Old Wanstead House, planted by
Sir Josiah Child.   This brings you to the main entrance

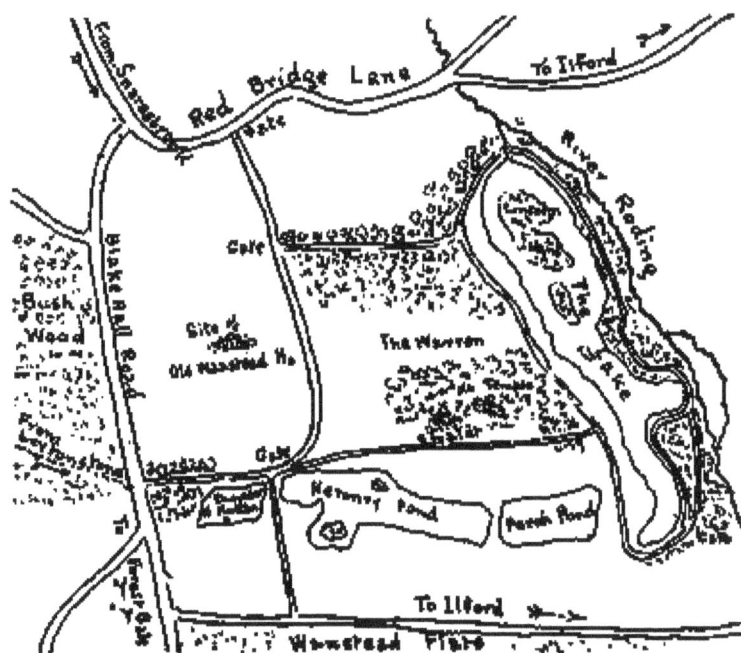

to the park in Blake Hall Road.   The same point is
reached by walking across Wanstead Flats—a dreary
name for a breezy common—from Forest Gate Station,
equally well served as regards train service from the City.
Yet another way is from Snaresbrook Station through
the village of Wanstead, past the Green and its broken-

down specimens of trees, down Redbridge Lane. A conspicuous notice board at the bend of the road will indicate where the entrance to the park may be made.

Supposing you had taken the Leytonstone or Forest Gate route : immediately through the gate a little copse has to be traversed ; then comes the first series of pools which is known as the Shoulder of Mutton Pond, devoted chiefly to the angler, but used indiscriminately by swimming dogs ; then the Heronry Pond (a misnomer, the herons are not here), is cut off half way by a fence and gate, which latter is open until certain stated hours, quite late enough for photographic purposes. The Perch Pond comes next, and is very pretty with its group of islands abundant with water-fowl, but the principal sheet of water is the long pond, or "the lake" as it is called. This is really charming ; the little islands dotted about its surface, the water-lilies lying on it, swans and water-fowl innumerable, its surroundings of tall trees all make it a picture.

It is best to "do" the lake systematically, beginning, say at the point where the path through the park joins that encircling the lake, close to the now dismantled grotto. This grotto, be it said, was the only relic of the old house and its belongings that passed into the hands of the Corporation when they purchased the property; a description of it would be difficult, but, roughly speaking, it was a kind of museum, a domed chamber laid out in fantastic devices with shells, stones, stalactites and crystals, stained glass and other materials, but not in any degree interesting from an artistic or antiquarian point of view. I remember, one raw November morning in

G

1884, seeing the still smouldering ruins caused by a fire which had completely gutted the grotto and originated no one knew how. It now serves as a boathouse—a roofless one—for the keeper's punt.

Pursue the path past the grotto until you come to a curious broken-down chestnut tree forming an arch over

the path. One of its branches seems to have become too heavy for the parent stem and has fallen into the water. Propped up, however, by a substantial wooden pillar, the broken branch is still growing and apparently flourishes, a source of wonder to most visitors to the park.

The gravel path still continues, and about a couple

of hundred yards further on is divided; the branch at
the left hand leads out towards Wanstead and Snaresbrook, but the other passes through a gate—if it be not
closed during breeding time, say from April to June—and
is one of the most delightful walks one can imagine.
Each season of the year has its own beauties of foliage or

The Brook Wanstead Park.

the want of it, and for wild woodland scenery I know of
no place to compare with it so near London. The river
Roding, here a brook, runs by the side of the path for
some distance, and at places affords some very pretty
glimpses, particularly further on. Now you come to a
bed of reeds stretching up into Lincoln island—you

cannot mistake it if you follow the map carefully—tall
trees surround it on each side and at the further end.
This is the "heronry," and you can almost always see at
least one or two of the long-necked, ungainly birds
perched up above their nests on the tallest trees; their
chattering and hoarse croaking is unmistakable, and is
most frequent when the herons begin to pair in the
Spring. A sudden noise or the report of a gun will send
them flying up into the air, wheeling sometimes so high
that they appear like swallows. Past the heronry and
Rook Island there is a splendid view of the ruined grotto
across the lake, and on the other side, if the water in the
brook be low—as it frequently is—it is worth while
descending to its bed by the brick wall to get a perfect
realisation of what a brook should be. Here the path
divides again, and it is of no consequence which branch
you take—they both join issue at another gate, the
counterpart of one you came in by, which is also locked
during the breeding season.

What strikes one so forcibly in Wanstead Park is the
unartificial nature of it, so strongly contrasting with
general ideas of London parks with their trimly-cut paths
and laid out flower beds. Here you have Nature at her
best, and it seems exactly as though some quiet country
seat had been picked up out of Kent or Sussex and
plumped down within a stone's throw, metaphorically
speaking, of the great city, to puzzle posterity how it
came there.

CHAPTER XVI.

THE VALLEY OF THE RODING.

THE RIVER RODING.—WOODFORD BRIDGE.—BUCKHURST
HILL TO LOUGHTON BY THE RIVER.—CHIGWELL:
THE "MAYPOLE" OF BARNABY RUDGE.—THE
CHURCH AND A CURIOUS MEMORIAL.—REMAINS OF
HAINHAULT FOREST.—LAMBORNE END AND CRAB-
TREE WOOD.—BACK TO ABRIDGE AND HOME FROM
THEYDON BOIS.

THE Roding is not a river of any importance—in fact,
few people seem to know of its existence at all; it has
its source or sources in a limited area south of Epping,
and joins the river Thames in unpleasant contiguity
to the outfall of the great sewer at Barking. Nevertheless, it has its points for photographic consideration; it
ran in the last chapter by the side of Wanstead Park,
and in this we will take it nearer to its source and com-
mence with Woodford Bridge. Here it is a very little
stream indeed, but it had volume enough to serve a water
wheel situated on the high road between Snaresbrook
and Woodford Bridge, in neighbourhood of George Lane
Station; the wheel itself is now pulled down, but the
stream is crossed by a wooden bridge carrying over a
footpath to Ilford hard by, and runs also parallel to the
road for some distance on the way to Woodford Bridge.
Buckhurst Hill is a better station from which to find the

Roding, and a walk from here to Chigwell, would be remunerative. To reach the river you come down the hill from the railway station and take the first turning on the left—a side street—which brings you into a new road cut between Buckhurst Hill and Chigwell, marked by a long stretch of white fencing; here you take the right

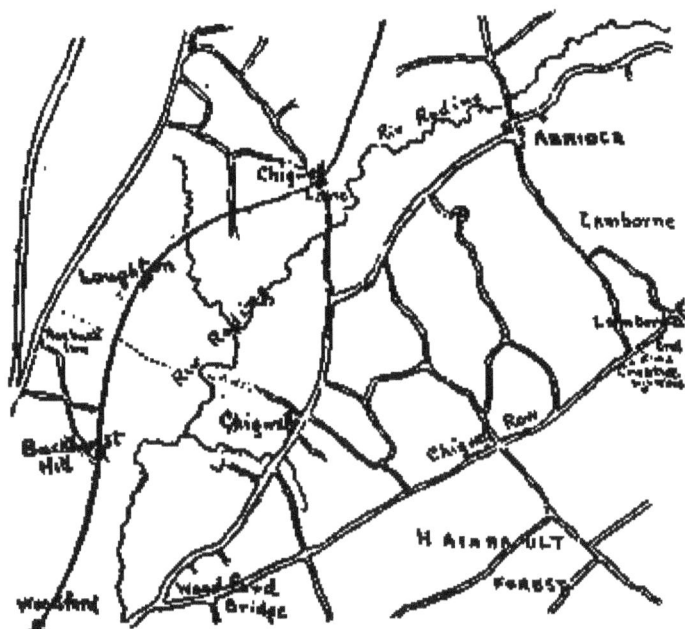

hand, and see across the road a wicket gate, through which over the first meadow you find the river, or rather a small arm of it, which joins the parent stream further on. It is only a tiny streamlet at this point, but a clump of gnarled and twisted willows bending over a bed of

reeds makes a fairly satisfactory picture; higher up at
the other end of the field is a wooden bridge painted
white, and the main river runs a very sinuous course
underneath it. The banks are well lined with shady
willows and other trees, one of which has fallen right
across the river and is still growing. The stream presents many pleasant corners between here and Loughton,
and though there is no well-defined footpath along it,
there seems to be no disposition on the part of the
riparian proprietors to treat the photographer as a tres-
passer. It is open, however, for no great distance,
although it may be followed almost to Loughton Bridge.

At the white bridge previously mentioned the path
continues up to Chigwell, and from its associations with
Dickens and "Barnaby Rudge" this village should not
be passed over. You will find opposite the quaint
wooden-spired church a very old-fashioned hostelry
known as the "King's Head," identified by Dickens's
letters to his friend Forster as the "Maypole," beyond
any reasonable doubt. There is a "Maypole" at
Chigwell Row, another village on the higher road, but
presumably it came into existence after Dickens had
made the name famous, and the reputation of the "King's
Head" has been in no way endangered by it. Be that
as it may, the "King's Head" is a delightful old inn—as
fine a specimen of what an inn should be within many
miles of London, but it is being rapidly modernised.
The city waiter in his swallow-tail looks out of place in
the grand old "Chester" room, as it is now called, in
contrast with the oak floors, oak ceiling, wainscotted walls
—suggestive of trap-doors—and its mullioned windows,

and most of the pilgrimages made to it seem to be more on account of the pigeon-pie, for which the inn is famous, than for its historical or literary reminiscences. The gardens belonging to it are very pretty, with their well-kept bowling-green and trim-cut hedges, and from the shady arbours you can follow the graceful outlines of the high-pitched roofs and massive chimney stacks. Flocks of pigeons, unmindful of their coming fate, flutter about almost tame enough to be caught by hand; truly this is a spot where laziness comes over one, but we must return to Woodford Bridge. It should be said before doing so, however, that the church opposite contains a very curious brass in the chancel floor to the memory of Samuel Harsnett, sometime Archbishop of York, the wording of which is so very extraordinary that I must be permitted to quote it; it runs: "Here lieth Samuel Harsnett, formerly vicar of this church. First the unworthy Bishop of Chichester, then the more unworthy Bishop of Norwich, at last the very unworthy Archbishop of York, who died on the 25th day of May in the year of our Lord 1631. Which very epitaph that most reverend prelate, out of his excessive humility, ordered by his will to be inscribed to his memory." One cannot help thinking that his grace—for he wrote it himself—used the wrong adjective, and meant "aggressive" humility.

Where the road forks at Woodford Bridge, the lower one leading to Chigwell and Abridge is that which we have just left, the upper runs to Chigwell Row and Lamborne End. This latter is equally interesting, for it traverses what was once Hainhault Forest before it was disafforested in 1851, when steam ploughs were set to work to root up the

magnificent oaks it contained, including the "Fairlop,"
which Gilpin says, in his "Forest Scenery," measured
36 feet in circumference at a yard from the ground,
dividing "into eleven vast arms, yet not in the horizontal
manner of an oak, but rather in that of a beech." The

On the Roding   Buckhurst H.U

only trace of Hainhault Forest now remaining is the little
piece of woodland lying between this upper road and
Barking Side, entered by a gate about half-a-mile from
Woodford Bridge, and another stretch up at Lamborne
End, now known as Crabtree Wood, both typical of
forest scenery and very little frequented.   On the way to
Lamborne you are able to look across the flat and level

expanse of Hainhault in its new estate as a Government
farm, and it is almost impossible to conceive that such a
dreary and monotonous tract was once the wildest of
smiling woodlands. At the "Beehive," at Lamborne
End, a lane turns sharply round to the left, a very pretty
lane leading to Abridge, but as there is nothing of interest
to be seen at the end of the journey, it is hardly worth
following; the only pretty view of Abridge is to be had
from the rising ground over which the road runs to
Theydon Bois, its nearest communication with the railway.

CHAPTER XVII.

BURNHAM-ON-CROUCH.

A New Country.—Billericay and Burnham-on-Crouch at Cheap Fares.—The Estuaries of the Roach, the Blackwater, and the Crouch.—Pictures of Marsh Land and River, and a View of Canewdon Church.—A Hilly Country: Essex Farmhouses and Cottages.—Rayleigh Mill and Runwell Church.—Maldon and the Blackwater.

I CONFESS that I should not have gone so far afield as the estuary of the Crouch had it not been for the enterprise of the Great Eastern Railway. Burnham must be more than forty miles out, but when the railway company will carry you there and back for eighteen pence on stated occasions, a visit is worth consideration.

The fact is, that the great tongue of land lying in the south-east corner of Essex—a tract of twenty miles broad and thirty long, roughly speaking—has only recently been brought into communication with the outside world, and surprise must have been felt on both sides as to mutual existence; a hardy, sober, industrious population, chiefly dependent on the oyster fisheries for their means of livelihood, are enabled to run up to the great city and see the Lord Mayor's Show, and the tired Londoner has the privilege of obtaining access to a part of the country as interesting as hitherto inaccessible.

The facts are these :—The G.E.R. run cheap half-day excursions to Billericay and Burnham-on-Crouch on Thursday and Saturday afternoons in the spring and summer, leaving London at about two o'clock, to return about 8.30; the fare is 1s. 6d., or 1s. 9d. second class

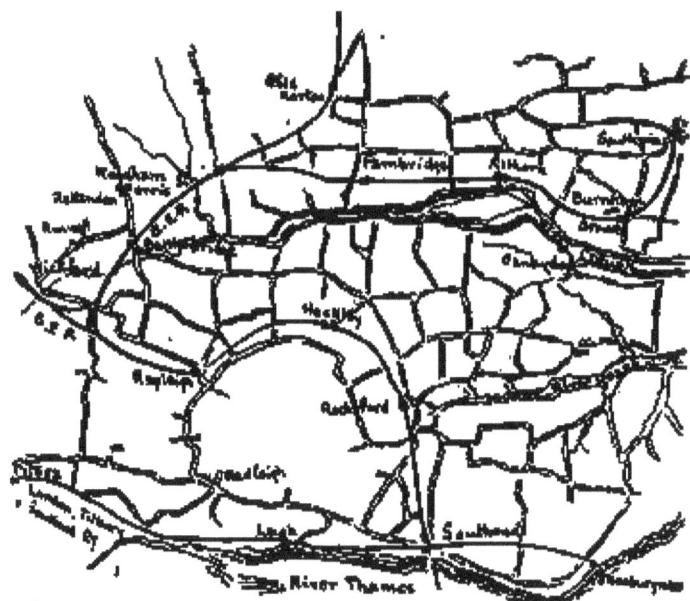

Scale 4 miles to the inch.

return.   On Mondays a whole-day excursion is run, and the return fare is 2s. 6d. The longer trip is to be recommended, and will be found all too short for photographic work.

I do not mean to say that the estuaries of the Crouch, the Blackwater, or the Roach, have any striking physical beauties; they have not, but for certain classes of study

they are all valuable. At low tide long stretches of mud
and "saltings" (as the great tracts of coarse grass land
submerged at high water are called), extend for miles and
miles broken here and there by a few stunted and wind-
bent trees, thrusting their arms into the river as though
to keep it well within bounds; long, low stone walls,
encrusted with sea-weed of a vivid green, vainly attempt
to stem the incoming tide; the river is full of craft of all
sorts and sizes from the oyster-dredger to the brown-
sailed smack and broad-beamed coble. The fisherman,
sturdy and weather-beaten, are studies too, and seem a
distinctive race, proud of their distinction. One cannot
see the beauties of such a place as Burnham-on-Crouch
at the outset; they are the outcome of long acquaintance
and close study, together with a *wish* to see them and the
bright side of things generally, which in photographic
work, as well as any other branch of art, is imperative to
real success.

The railway station at Burnham lies about a mile
north of the town, if it may be called a town, consisting
as it does of one main street and a promenade—neither
does the latter lie at the mouth of the river, but about
five or six miles inland; nevertheless, it is pleasant to sit
on the "hard," or on an upturned coble, looking up the
river (which is here from five to six hundred yards wide)
in the direction of Battlesbridge, and to watch the setting
sun light up the weather-beaten tower of Canewdon
Church, two miles away, a landmark for the country side,
the river gleaming like polished silver and the trees along
its banks bathed in a mist of golden sunshine.

As to the surroundings of Burnham, I have recently

done by road the country lying between the Blackwater
and the Crouch, and found it more interesting than I had
believed to be possible. Essex is reputed a flat country,
but the flatness does not lie in this quarter; in touring
you are perpetually climbing hills and descending into
the valleys; you mount the ridges of Rettendon or
Runwell and look across the estuary of the Crouch, or

Burnham on Crouch

from Cold Norton and Althorne you see the valley of the
Blackwater with Maldon in the distance, a "give-and-
take" country in which the latter feature is prominent.
What cannot fail to strike you in travelling these roads is
the picturesque appearance of Essex farmhouses —the
gables, the red-tiled moss-grown roofs and the immense
haystacks ; the bits of wayside common land watered by
trickling streams; the cornfields and the poppies; cottages
of wood, brick and plaster, seemingly unique in design

and wanting in nothing from an artistic standpoint.
Added to this the simplicity and good nature of the
inhabitants, who invariably give you "good-day" and a
salute when you come across them. The railway, it is
to be feared, will alter all this in time, but there is a
favourable opportunity just at present for the photographer who can afford to spend a week or a fortnight
amongst them. I cannot particularise any localities—
they all have special features of their own. Rayleigh
with its mill crowning the rising ground, Runwell for its
pretty church and cottages, Woodham Ferris for its
situation, Fambridge for its river and ferry—all these
villages deserve to be known and visited by the photographer. The new line, branching at Wickford, goes in
one direction as far as Southminster, and in the other to
Southend at the mouth of the river Thames, while at
Woodham Ferris still another goes to Maldon, hitherto
reached by way of Witham, on the Colchester line.
Maldon is a place of some importance as a fishing centre,
standing on the broad estuary of the Blackwater, about
fifteen miles from the open sea.

A map is very useful in these parts, owing to a want
of inter-communication. There is no lack of finger-
posts, but there are so many bye-roads and cross-roads
to different places, some of which arrive at their destination by very round-about methods, that a "reduced
ordnance" will save perhaps many miles of weary travel.
A handy little guide, with an accompanying map, is Percy
Lindley's "New Holidays in Essex" (6d.) ; it particu-
larises the new line of the Great Eastern, and will give a
good general idea of the country from points of view
other than photographic.

There are many people who know nothing of the existence of a sea wall and the necessary function it performs, round the coasts of Essex in particular. What is the wall? Well, really it is nothing more than a high embankment of solid mud, strengthened in some places by stakes; in some by stones loosely placed together, according to the exigencies of the locality in which it finds itself. Of masonry, as we understand it in speaking of a wall, there is none.

The sea wall has a fascination for me, whether it skirts the Tilbury Marshes or the dreary shores of Canvey Island, and I have lately followed it in its wilder phases at the estuaries of the Colne, the Crouch, and the Blackwater, unknown rivers of England. Between breezy, briny Maldon and the mouth of the Blackwater there are thousands of acres of reclaimed farms lying ten feet lower than the water in the estuary at high tide, and here the wall is broad and solid, for one could understand what havoc there would be if the embankment gave way, and what miles and miles of inland country would be devastated.

You may walk along the wall for hours together in this country without meeting a living soul, but from the land side you may continually be putting up a covey of partridges, and watch on the saltings the heron fishing, stolid and motionless, the flocks of curlews that come piping overhead, the snipe whistling as he rises flying in the wind's eye, the sandpiper and the wild duck. A lover of nature can never find himself alone.

The wall is densely covered with coarse sand-grass, and sand-grass is a land winner of some importance; it

grows to some height, and its light straw colour shaken
up by the breeze gives very beautiful effects of light and
shade as you traverse the sinuous course of the embankment. With the sun full upon it you see on the ground
glass of the camera a curious sugar-cone structure formed
by the rustling grass as it rises and falls on either side.

At low tide—and it is at low tide that the wall is at
its best—you will notice on the seaward side great
stretches of apparently solid earth cut with deep channels, dividing it into irregular and fantastic shapes. This
earth has a top covering of coarse vegetation, seawort,
and a beautiful flowering marsh plant, the name of which
I think is artemisia, and is entirely submerged at each
succeeding tide. Hence the term "saltings," a generic
term in use all round the coast. To appreciate this
magnificent desolation one should visit the "saltings" in
the depth of winter, when the sky is dark and lowering,
and the slanting beams of the sun's search lights break
through the clouds. The centre beam casts a strange
shaft of light that comes dancing over the river until it
reaches the treacherous, slimy ooze, gleaming like
myriads of liquid diamonds—can mud ever be made to
look more beautiful?—and flashes on the shallow pools
left by the last tide, or gradually filling with the incoming
water.

The sun is very low down throughout the day in
winter, and to photograph successfully against the light
great care has to be exercised. These slanting rays, too,
must be very strongly marked against a dark sky, or you
get no evidence of them in the developed plate. I have
experienced this many times with disappointment. They

H

are rarely stationary for more than a second or two, and the clouds break up often before you can get your camera into position; the skies are ever changing.

Striking the wall from Maldon, one may follow the mouth of the Blackwater for many miles in either direction. It winds a great deal, but for photographic purposes it matters very little, so that you calculate times and distances from your base of operations. Fog is an element of danger at all times, and these low-lying lands are particularly susceptible to it, but in the event of its coming on, it is as well to strike the road across the fields as soon as possible, to find that safety on its hard surface that is not to be met with on the slippery banks of the sea wall.

If you take an ordnance map—a useful accompaniment to the outdoor photographer in a strange country— you will find, coming from Maldon, a road to Goldhanger clearly marked thereon. It runs through Heybridge, and about half a mile beyond that village abuts on the wall, where a windmill towers above it. From this point it is all plain sailing. Here is an old decoy, disused and reed-grown. It was originally a large pool, with narrowing tunnels, made of oziers, facing the different winds that blow, and cunningly contrived with many angles to the end that the unwary ducks following the grain scattered at the mouth eventually found themselves in a bag net at the further extremity. I came across an old weather-beaten shepherd, who had been watching my strategic movements on the wall for some time with curiosity—the only living soul I had met for five hours—and got into conversation with him. He remembered, not so very

long ago, when this old 'coy were a thriving concern
before wild duck got so scarce. You mighn't think so to
look at it now, but a bag of three hundred in a week was
common enough. From wild-duck the talk got on to
smugglers and smuggling, and I fancy I detected a
twinkle in his eye when he assured me that since the
coastguard was established at the point you never. heard
of " moonlight "—another name for Hollands—and Tip-
tree Heath was only a memory. What wicked old times
they were!

Then, as you don't get much chance of conversation
on the sea wall, I asked him where his sheep were, and
he said he had a matter of three hundred across the
" mairsh," some of 'em due to lamb in a fortnight, some
not before the first week in February—white-faced Kents
they were, and as clean a looking lot of sheep as ever I
saw. So we went together across the disused decoy to
the " coy fairm " to look at them. What simple-hearted,
honest souls these Essex shepherds are.

He rightly called this field over which we are
travelling a " mairsh," for it is very bountifully watered
by little shallow pools, round which one picks a careful
way. Here a hare springs up from her furrow, and
startled, runs skirting the staithe to take refuge through
the huddled mass of sheep. " Mad as a marsh hare,"
surely the proverb ran at first, for there are none so wild,
and yet, had we kept on our course without taking any
notice of her, and out of the direct track, we might have
passed within a yard without starting her. Poachers
as well as gamekeepers know this fact in natural history,
and having " spotted " a hare in her form, keep on

apparently not noticing anything, until suddenly down drops a heavy stick in the grass, and a dead hare finds its way into a capacious pocket that extends right round the back of a fustian jacket.

But this is not photography. Let us return to business and the sea wall. What has happened since we left it to look at a flock of sheep a hundred yards away ? All the landmarks are gone, not a bit of " salting " visible, and the water which was eight or nine feet below the shelving bank is now within a foot of the top. Photography is gone for the day, unless one can catch those beautiful effects of sea and sky from the vantage ground of the wall. It is late in the afternoon, and the sun, glowing like a crimson ball, is sinking behind the long row of fishermen's houses that stretches out from Maldon to receive its dying rays. The rippling lights leave the water, and the hull of a derelict barge that a moment ago was glistening with fairy lights becomes a deep grey, then a black shadow. The day is done.

CHAPTER XVIII.

SOUTHEND-ON-SEA, LEIGH &c.

SOUTHEND-ON-SEA.—AN OLD FISHING VILLAGE AND ITS
INDUSTRY.—LEIGH AND CANVEY ISLAND.—DUTCH
COTTAGES, DYKES AND MARSHY LAND.—CATTLE
STUDIES ON THE ISLAND.—HADLEIGH CASTLE AND
ITS CROWNING POSITION; BEST VIEW FROM THE
SLOPE ON THE THAMES SIDE.—A FLOCK OF SHEEP.
—SOUTH BENFLEET AND ITS FERRY.—ONE OF THE
PRETTIEST CHURCHES IN ESSEX.

WHILE we are in the south-east corner of Essex,
Southend and its immediate surroundings have
claims on the photographer that are not to be passed
over. Southend is growing considerably and rapidly,
owing possibly to a new line of railway and increased
accommodation, and for a summer holiday is to be
strongly recommended, not so much for its own sake,
perhaps, as for its convenience as a central point from
which to do the neighbouring country.

All along the esplanade or sea-wall, stretching for
more than two miles, is as favourable a place for sky
studies as any I know; you can look across four or five
miles of open river on the Kentish hills and follow the
coast outline right round by Sheerness and the Medway,
or trace the long white track of the P. and O. steamer

ontward or homeward bound, the brown-sailed lugger
and the grimy collier making for the open sea.

But it is not Southend itself that will find the photographer—the earnest photographer, not the man of snap-
shots—that abundance of work which is to be found at
Leigh, at Hadleigh, or Benfleet and Canvey Island.

Scale 4 miles to the inch

Following the sea-wall westward, a walk of three
miles, a delightful walk, too, will bring you to the strand
at Leigh, an old-world fishing village, full of interest as
its port is full of shipping. They are building at Leigh,
too, but not to the same extent as its larger rival, and it
will probably be some years before they have altered its
characteristic quaintness. You get a view of Leigh
church and its surrounding trees perched up on the hill-
top long before you reach the village (it is a curious fact

that in this part of Essex, wherever a hill is to be found, almost invariably a church is made to cap it); and as the eye drops down lower it takes in the little port at its foot, then long promontories jutting out one after another until they seem to be lost in the haze of the great city miles and miles beyond.

The special industry of Leigh seems to be the catching and cooking of cockles. Along the beach you will find numerous wooden cabins or sheds, with a large open window at the one end and a heap of empty cockle shells beneath and around it. Watching the fishermen and their primitive methods, you will see how these molluscs are prepared for the London market, and—in parenthesis —you will vow that if you have never eaten cockles before, nothing is likely to tempt you in the immediate future. They are thrust into a boiling cauldron fitted inside the cabin, fished out after the regulation time has elapsed with a net, and riddled in an ordinary sieve; the cockle subjected to this treatment, open-mouthed with astonishment or long boiling, drops his interior through the sieve on to a board below—a board that has served for ages— and his empty shell is flung aside to be ground with asphalt or in a crushed state to serve as garden "gravel."

This digression has not much to do with photography, perhaps, but it is the every-day life of Leigh, and deserves mention if only on that account. Crossing the railway at the station, take the first turning on the left—a quiet bye-street—you re-cross two or three hundred yards lower down where you see a gate opening on the line. This brings you to the beach, and a path runs parallel with the railway for some distance, affording through its

whole length far-reaching stretches of low marsh land on
Canvey Island on the one hand, and hilly tracts of Essex
upland on the other.

Canvey Island is, roughly speaking, five miles in
length, and three miles wide at its broadest part; its
area of 6,000 acres was reclaimed from the sea by an

Hadleigh Castle

enterprising Dutchman in the seventeenth century, who
claimed and received one-third of it for his trouble. This
accounts for the Dutch cottages dotted over its surface,
the dykes that intersect it in all directions, and the Dutch
customs that are still said to survive in the island.
These dykes stretch far up inland, and the river creeks
harbour little craft of all kinds, as well as an old derelict
hulk, now on its beam ends. Cattle abound on the rough

pasturage that nature has provided for them, and the
solitary heron has set up fishing on his own account;
truly, Canvey Island affords many effective studies of
sea and sky, expanses of wild marshy land and gleaming
river, never absolutely desolate, but impressive in a
marked degree. The regular ford and ferry is at South
Benfleet, four miles from Leigh, and the railway station
is within a few yards of it. You will have noticed on the
way from the latter place, Hadleigh Castle, lying two
miles west of it; there is a footpath after crossing the
line again up the hill side, and with the exception of a
little piece of bog—dry enough in good weather—it is an
easy one to follow. With such a landmark as Hadleigh
Castle in front of you there is no going astray, in spite of
the path becoming at places only a sheep track; you see
the two grim and grey towers from afar, and, to use a
homely metaphor, follow your nose. All that remains of
the castle is to be found in these two towers, crumbling and
weather-beaten, a fragment of gateway, and most of the
wall foundations which crop up through the sparse grass
and undergrowth, struggling through the thick masses of
ivy and briar in a well-defined outline. The disadvantage,
from a photographic point of view, is that the remains
are too scattered to obtain a representative picture of the
whole; the castle stands on the crest of a hill, and is
surrounded by trees and bushes. The most complete
survey seems to be from the slope on the Thames side;
from here you command a full view of the principal
tower and a glimpse of the one facing north, with ivy-
grown walls and undergrowth in the foreground. Spring
time is a very favourable season, as the condition of the

trees allows a fairly uninterrupted view of the whole castle, and there is the added incentive of a flock of sheep and lambs—the tamest flock I ever came across—apparently taking a lively interest, and not too lively, in your work, generally to be found in the castle grounds. Hadleigh Castle forms the subject of one of Constable's best known pictures.

Since this was written General Booth has purchased the Hadleigh farms together with the ruins thereunto appertaining, but I do not think it need make any difference to photographic work.

From Hadleigh to Benfleet is still another two miles, and the church is one of the prettiest in the county with its massive stone tower, brick buttresses, and wooden porch; its picturesque graveyard and surrounding elms in which the rooks have established a colony of their own. To sum up, all the places I have mentioned are well within a day's work, and the London, Tilbury, and Southend Railway has stations at Benfleet, Leigh, and Southend. A glance at the map will show how near are all the points enumerated.

CHAPTER XIX.

IN A FLAT COUNTRY

CANVEY ISLAND AGAIN.

THE appreciation of a flat, marshy country must be to most photographers an acquired taste, but, having acquired it, the love for it grows upon one speedily and surely in such measure that no mere prettiness can compensate for its value in picture-making. This feeling has been accentuated in my own case by the study of a photograph by one of the leading workers, which should have made its appearance in the present exhibition at Pall Mall and did not—a picture of a wild bit of "Salting," with its coarse grass and brackish pools, its damp atmosphere, the distant glimpse of the river beyond, and overhead a threatening sky through which the beams of light struggle in slanting rays to throw a weird light on the centre of the picture.

Have you seen Henry Moore's "Clearness after Rain?" What is it, after all? A stretch of sea and sky without artificial embellishment of any kind, yet such a picture as compels you to *feel* that it has been raining, and the cold wind is blowing in your face. A mere transcript of nature—a poll-parrot phrase it is the fashion of shallow photographic critics to use by way of expressing their contempt of certain works—describes it accurately; nature, pure and unalloyed, graphically demonstrated, demanding a master mind to transcribe, artistic instinct to seize at the right moment, and last, but not least, the necessary technical skill to execute. As it is in Henry Moore's painting, so it is in this photograph of "Saltings" I have just mentioned; full of breadth and feeling and—what is rare in photographic productions—atmosphere, it makes one doubt whether there are art limitations in photography after all.

And the country that yields such pictures is here at our very doors almost, unsought and uncared for except by the very few. If you take a map you can trace round the coast-line of some part of Kent, Essex and Norfolk, huge tracts of this flat country, which is gradually being left bare by the sea, eager in its desire to undermine the Yorkshire cliffs. Canvey Island at the mouth of the Thames is a good example of it, although history says it was reclaimed from the river by Dutch settlers—aided by nature probably it was. Then there is the region of the Norfolk Broads, which has still its hold on Dr. Emerson, and can find a life's work for many more.

It was Canvey Island I was thinking of more particularly, possibly because it is nearer London than any of the others, and it seems of equal value regarded photographically. Benfleet is the station for it, and although it is only divided from the mainland by a creek of varying width there is no other place of crossing. At low tide there is a passable ford with a hard gravel bed, all the more remarkable because, on each side of it, as far as the eye can reach along the creek, there is nothing to be seen bordering the banks but a thick slimy mud, known locally as "ooze." A plank here and a stone there afford the foot passenger safe, if not over clean, means of transit. On the other side the first thing that strikes one is the absolute dreariness of its aspect, but as one gets further inland there are signs of habitation, a farm here, a shed there, and right across the island a little iron church. There are dykes, reed and weed-grown, to separate the different pastures; and high embankments, called in Norfolk "rands," to keep out any possible further encroachment of the river, giving scanty shelter to the herds of cattle. The surface of the "Saltings"—and the term "Saltings" is applied not only to those portions of the land which are covered with salt water at high tide, but to those also which have been reclaimed or partially reclaimed, but still contain their saline properties—is broken up by little pools of brackish water, sometimes standing singly, sometimes connecting with each other and the dykes by narrow and tortuous channels. You will see on some of these pools square sods of tufted grass placed at regular distances from each other and for a moment you are puzzled to know the meaning of it. Then it dawns upon you that this must be the Canvey method of land-winning. The mud accumulates on these little patches of turf and keeps on accumulating; seeds fall and the grass roots bind it together, weaving a network of fibres that seasons as they roll will gradually fill up with solid earth. Can one help feeling admiration—I had almost said reverence—for a labour such as this, working not for to-day or to-morrow even, but for results that will only benefit an after generation?

The centre of the island is well under cultivation, and the pasturage is good, but on the miles and miles of

"Saltings" nothing flourishes but coarse marsh grass and a peculiar flowering weed—the name of which I have forgotten, if I ever knew it—that is very valuable for foregrounds, if nothing else. On these cheerless expanses the summer sun throws down its blistering rays, scorching the marsh grass like rags in a tinder box, and the cutting March wind with pitiless rain behind it bites one to the very marrow as it tears along from the sea. Yet the cattle seem to thrive under stress of wind and weather—perhaps they are born to it as the ponies are to blindness in the Russian salt mines—and for animal studies of a particular kind, flat country has a distinct value of its own. It certainly sharpens the wits, and makes demand on the photographer's powers of composition to make satisfactory pictures out of insufficient materials, and for this reason alone, if for nothing else, it is to be recommended.

To get back to Canvey Island, one road runs right across it to the Sluice House and to Holehaven (where, if ill-natured gossip is to be believed, there is still a strain on the watchfulness of the Excise officers), while various narrower ones branch from it here and there, generally leading to a farmhouse, and stopping there. You may meet with weatherbeaten shepherds, with corduroys roped at the knees, and a coarse canvas covering the shoulders, leading their flocks to fresh fields and pastures new, willing to exchange remarks on the weather, and to put themselves and their charges into position for the love of the thing, and a desire to encourage photography. If you look back towards Benfleet you will see what a pretty little village it is, with its ivy-towered church and broken up red roofs, and across the "Saltings" and the creek a pinhole exposure might be made to do it justice. Away to the right are the ruins of Hadleigh Castle, and beyond that is the fishing village of Leigh, and the smoke of Southend just visible in the distance. Across the island the Kentish hills are faintly outlined, and the thin wisp of black smoke from a passing steamer draws a dividing line between them and Canvey.

How the wind blows in chill October! One can hardly keep a foothold on the muddy rands burdened with a heavy camera. Darkness seems to come on all at once, and the mists rise damp and cold from the marsh. One can almost realise the full meaning of solitude here when night has fallen—no sound but the cry of the snipe or the seagull, for the sea is miles away, and the water in the creek without a ripple. A feeble, flickering light from solitary farms far away is the only sign of habitation. The tide has risen, and the road from the ford is lost in it, as the incoming water stretches up for three or four hundred yards into the island. From the higher level of the footpath it gleams strangely grey in the half light, cold and treacherous. But there is the sound of oars at the ferry, for the boatman has seen two black figures against the sky—mad photographers surely—for he thinks they can never make pictures on Canvey!

CHAPTER XX.

BENFLEET, STANFORD-LE-HOPE, LOW STREET, &C.

BENFLEET CHURCH: THE BEST VIEW OF THE VILLAGE.
—PITSEA AND BOWERS GIFFORD TO STANFORD-LE-
HOPE—THE LANGDON HILLS—LOW STREET: A BIT
OF COMMON LAND AND ITS SITUATION—MARSHY
LAND: ALONG BY THE DYKES TO THE SEA WALL.

BENFLEET Church was our limit in the last chapter, but I omitted to mention that a good general view of the village and church may be had from the other side of Benfleet Creek. This is reached by a footpath through the churchyard, crossing the creek at its narrow end by a wooden plank. There is a tolerably broad expanse of water here when the tide is up, but at low water the "ooze" is not objectionable in any way; its lines are well broken by various channels, and the fisherman's smack aground in the mud makes a very presentable foreground to the picture of the church and cottages on the other side of the river. The farmhouse almost adjoining the Church is "Sweetbriar Farm"; nor does it belie its name, but with its thatched roofs and rookery may be taken as a typical farmstead of south-east Essex.

In leaving Benfleet I cannot help thinking I have done Canvey Island scant justice; it needs closer study

by photographers from the ford and ferry at Benfleet to
the "sluice" across at the other side of the island.
Usually there is a very picturesque barge laden with
rushes or something of that nature close by the ferry,
and it makes a very effective picture with a figure or two
coming from Canvey Island, finding the cleanest or driest

Scale 4 miles to the inch

path across the river bed. Reviewing the line of railway
on its return journey to London I shall only be able to
give the salient features of each district; for it must be
remembered that most of these stations are some
distance apart, and walking from point to point is
difficult owing to the dykes. The farmers are usually
good-natured enough to let you trespass where you will

I

in search of the beautiful, but short cuts on marshy land
are rarely productive of anything but bad language and
wet boots.

Pitsea is the next station nearer London, but about a
mile before reaching it you see lying in a hollow close to
the line a very curious old church built of grey stone,
with a tiny wooden steeple and porch; this is Bowers
Gifford, and a very pretty little church it is, surrounded
by trees and a sparsely populated graveyard. They used
to build their churches in the hollows formerly, sheltered
from the north-east winds, but now they put them on
the highest hills, and plant trees all round them, as they
have done at Pitsea, Vange, Langdon Hills, and any
number of places in this neighbourhood. There is
nothing very much at Pitsea if we leave out Vange Creek
and the marshy lands on its borders, its farms, and sheep
and cattle, but from Stanford-le-Hope, the next station,
you can reach Horndon-on-the-Hill, an upland parish
with a quaint, wooden-spired church and a windmill, and
from there find the Langdon Hills beyond.

To prevent misapprehension I should explain that
Langdon or Laindon Hills is really a small village standing on the crown of the hills of the same name; its nearest station is *Laindon*, one mile north, and its chief
recommendation lies in the beautiful and extensive views
you can obtain of a lower valley of the Thames from
different points on its summit. For a pedestrian it would
be a charming place to visit, but for the photographer—
well! I have not yet been able to decide whether the
view of the old church—a harmony in brick red and green
—lying at the foot of the hill on the Romford side, is

worth such a long and telling climb one has to incur to
find it. The new church at the top of the hill was not
built for beauty—a heavy, solid stone structure visible all
over the county—but it is very picturesquely situated,
and the woods intersected by paths that run down the
hillside are really very charming.

For a certain class of picture afforded by flat country, ·

At Low Street.

winding brooks and rushy dykes, flocks of sheep and
lambs, and herds of cattle, commend me to the village
of Low Street, lying between Tilbury on the one hand
and Stanford on the other. From the end of the station
platform you can see a bit of common land with pools of

water, clumps of willows and sedge, and a couple of
cottages at the side, a picture that would inspire Leader,
if he could see it. It lies really a few yards beyond the
platform, but may be reached equally well by the road ;
to explain its situation exactly, coming from London you
would cross the line and take the road for about two or
three hundred yards, bending round to the left; then
you come across this bit of common, with a pumping
station at the further end. Seen from here, no photographer would waste a plate upon it—there is nothing
pretty about it at all; but look at it from the other side
and you will be surprised at its changed aspect, and if
your dark slides only hold six plates, five of them are
bound to go. To return to the high road. There is a
gravel pit on the other side, but it has nothing particular
to recommend it. Follow the lane again, still to the left,
until you reach a farmhouse—there will be no difficulty in
finding it, as it stands by itself, and has some fine
thatched barns. At this point the lane turns abruptly to
the right by the side of a duck-pond, it gets deeply rutted
further on, and changes its character by degrees from a
country lane to a footpath, eventually becoming a sheep-
track and leading down to the sea wall below Mucking.
Great clumps of teasels shoot up from the marshy land,
groups of willows and other trees break up the flatness
of it, but the special feature to be noticed is the reedy,
winding dyke whose banks are covered with countless
long-woolled sheep and cattle. The marsh is innocent
of hedges, but the dykes—none of them cut with mathe-
matical precision—intersect it everywhere. Farms are
dotted over the landscape, and the eye travels from the

smoke of Tilbury and Gravesend to the Kentish hills beyond, with a mile of busy river, alive with craft of every kind, intervening.

This station (Low Street) has been a favourite resort of my own for many months, and there is scarcely a shepherd or a flock of sheep belonging to the farms in the vicinity whose acquaintance I have not cultivated. It was on a gate in the lane I have mentioned above that I jotted down notes for the following article; which appeared in *Photography*, September 24th, 1891.

CHAPTER XXI.

SITTING ON A GATE.

I HAD finished my work for the day—a long sultry September day it was, that rightly belonged to the previous month, its heat was so intense—the grey haze had never risen high enough above the meadows to clear the tree tops, and the cattle lay about in groups under the elms, seeking shelter, not from the sun, for his rays were hardly visible, but from the all-pervading sultriness which covered the land as with a garment. Every living thing seemed to lack the power of exertion; the birds were silent, as they usually are in the late autumn, even the flies had ceased to annoy an already overburdened photographer. As the afternoon wore on I found myself sitting on a gate—an old friend of many seasons—and reflecting lazily on the chances of a good day's work or otherwise, and came to the conclusion that the latter was most probable, taking into consideration the fact that all the exposures were made with a shutter.

It is just an ordinary five-barred gate, with nothing in particular to distinguish it from other gates except its charming situation—such a gate as Richard Jefferies would have loved to sit upon reflective. One cannot help thinking what a good photographer he would have made. A lover of nature in all her moods, his observa-

tion let no detail escape him; it is true he stopped down
every picture of his mind's eye to focus 32, and yet he
belonged to the naturalistic school, if any man can
claim a place in it, an exponent of it whom no out-door
photographer should feel ashamed to follow.

The late afternoon wears on, and a gentle breeze has
sprung up, clearing away the thick mistiness with a
suddenness that is incredible as it is refreshing. The
sough in the trees might be mistaken for the sound of
the sea with its regular rise and fall, the tall reeds that
border the dyke shiver in it after the overpowering burden
and heat of the day. Up the gravel bank on the other
side of the lane the convolvulus has forced its way
through briar and blackberry, striving with the wood-
bine's last flower to reach the higher life, the hemlock at
its foot has lost its whiteness, the bracken its garish
green, and one, involuntarily enough, perhaps, contrasts
the crudities of early summer with the mellow toning
down of glorious September.

Within a yard or two of the gate there is a sheep
wash cut out of the solid earth, like a grave almost,
lined like a family vault with brick and cement, and
lying in the bottom of it the slimy green water gives
out a faint indescribable odour as of centuries of sheep
washing. Then one muses on the running streams of
Leicestershire and Hampshire, and the picturesque
aspect of sheep life from a photographic standpoint; how
the unkempt, uncleanly sheep are bundled into clear
water and subjected to forcible ablution by the strong
hands that will afterwards shear the well washed wool.
Then the thoughts wander to Wiltshire Downs and the

picturesque shepherds of Salisbury Plain, the ram fair at Warminster, and back again in the flash that annihilates all distance to the salt marshes, here close at hand, where the most curious figurehead of all sheep studies is to be found in the old master shepherd, his horse and dog, who know all his methodical movements on the chequered board of a sheep farm, as well as they do each dyke in the marshy land and the gate across it that connects it with the field beyond.

Longfellow must have known my old shepherd, for does he not say in "Evangeline," "Then came the shepherd back with his bleating flocks from the sea-side, where was their favourite pasture. Behind them followed the watch dog."

Sheep are creatures of habit, too. You see them waiting here at a given gate and a given hour, not a straggler missing, and you wonder what it is has brought them so together, until you discover away over the marshes the old weather-beaten shepherd mounted on his shaggy long-tailed cob, leisurely making his way towards them and you. They know and he knows that the flock has been long enough on the "Saltings" to do them good. Your Wiltshire farmer will tell you that when you see a sheep drinking you may know that he is likely to die soon, the fever is already on him (and it is surprising what little will kill a sheep), yet here, when they come from the "Saltings," you may see them drinking in dozens, and the shepherd showing no concern.

Watch the old man how he dismounts from his horse, giving no care to the tying of him up, but opening the gate just wide enough to let through the flock in single file, how he counts them with crook in hand. There is no unseemly rush as he tells them off one by one until the tale is complete, apostrophising one, admonishing another, as though they were children; they spread across the plain, looking like little white dots on a great expanse of green and brown. Then he looks at his dog snatching a brief moment with his nose in his forepaws beside him, and the look is sufficient to tell him that the day's work is finished; your eyes follow the retreating figures, tuckered smock and battered hat, shaggy horse and tailless dog, until they are lost in the distance. Sturdy and independent he is, yet courteous as a gentleman should be. I remember in the early days of my work on an Essex sheep farm, I gave him on one occasion a lot of unnecessary trouble in the pursuit of picture making, and wound up with the mistake of offering him money by way of recompense. I have never done it since that day, but he is as proud of some photographs of mine in which he and his sheep figure as I should be if they had taken all the medals in the country. His hut lies at the bottom of the lane here, not at the next turn of the deeply-rutted road, across which the long shadows of bordering elms have broken up the sunlight, nor at the next perhaps, but where the lane is merged into marshland through the gate that divides the two, where the teasels grow in great clumps on the banks of the dyke, and the sheep-tracks begin to draw crooked lines across the flat country; a solitary house stands on the marsh. In the hot weather the sun pours down its pitiless rays on the parched soil, and the bitter east wind of February and March searches out every nook and corner, making the sheep huddle together for warmth. There is no shelter from either.

How recollection runs riot when sitting on a gate; the drowsy hum of the bee making straight flight for home, the mellow tinkle of the cow-bells as their wearers return from milking, have a soothing effect on a restless mind when coupled with tobacco.   But the sun sinks behind the gravel pit, and the afterglow is of short duration. the white mists rise from the fields and throw a veil over the dykes, the day is finished and there is silence.

CHAPTER XXII.

BRENTWOOD, WEALD PARK, THORNDON PARK, &c.

BRENTWOOD.— A PRETTY FARM.— SOUTH WEALD CHURCH AND WEALD PARK.— STUDIES OF RED AND FALLOW DEER.— THROUGH THE PARK TO PILGRIM'S HATCH.—FROM SHENFIELD COMMON TO WARLEY GAP: A VIEW OVER THE "GAP."— WARLEY COMMON AND THE PICTURES ON IT.—A RUSH-GROWN POOL AT THE FOOT OF THORNDON.— CHILDERDITCH STREET.— ACROSS THE PARK TO HERONGATE.— BACK FROM BRENTWOOD OR FROM EAST HORNDON.

EIGHTEEN miles out from London lies Brentwood, in the midst of scenery typical of Essex, yet altogether distinct from that phase of it we have recently seen in the lowlands of the Thames valley and the estuaries of the Crouch and Roach. This part of the county is hilly and well wooded; fertile also to a degree and abounding in choice "bits" for photographic consideration.

You will notice just before reaching Brentwood, at the left hand in the direction in which the train is travelling, a very pretty farmstead lying in the hollow with a little piece of water bordered with willows in front of it; from

the railway it composes very well, and there should be no
difficulty in finding it when you reach the station.   The
wooded height beyond it is South Weald, with its church
tower peeping out from the trees.   To reach this point,
make your way up the hill to the High Street, half a mile
from the railway, and enquire for Weald Lane, which is

Scale 1 inch to the mile .

a narrow turning on the left; this found, it is a straight,
or, more correctly speaking, a very crooked lane to South
Weald up hill and down dale.   A mile and a half out
from Brentwood you reach the wooden fence which
bounds Weald Park, and standing at the corner of a lane
at the right, leading to Pilgrim's Hatch, is a clump of

pollarded beeches inside the park that reminds one of those in Epping Forest or the still grander Burnham Beeches. The public footpath is about a quarter of a mile further on the road you have followed from Brentwood, and is to be found a hundred yards south of the church; before turning in the wicket gate to the park it will be well to have a look at the Church, if only to see how effectively restoration has spoiled it photographically.

Weald Park suffers under a great disadvantage in that notice boards are posted in prominent places warning the public to keep to the footpaths, and by far the greater portion of it is inaccessible. There is really only one thoroughfare through it, which connects South Weald with the pretty little village of Pilgrim's Hatch before mentioned; but this is worth following to the end, two miles away, as photographic opportunities are continually presenting themselves.

Not only are there some really fine studies of trees, but there are large herds of fallow and the more uncommon red deer to be met with, and also some curious specimens of long black-haired sheep with incurved horns. Beyond the lake and up the next hillside is a favourite spot of the red deer—noble looking animals they are, too—and though they are on the wrong side of the iron fencing they are fairly tame, and may be brought within the range of practical photography. If one only had the right to wander here at will, good work might be done with patient tactics, but it is the disadvantage I spoke of, that one has really to depend so much on blind chance and wait one's opportunity. An earnest worker will think it worth the risk.

Brentwood is rich in parks. Thorndon (Lord Petre's)
is one of the largest, if not the largest, in Essex. This
lies south, and there are one or two ways of reaching it,
the prettiest being by way of Shenfield Common and
Warley Gap. Supposing you have come from London :
once outside the station you would strike diagonally

Thorndon Park

across to the right through a wicket-gate, and then take
the first to the right again up a private road, past the
police-station. In half a mile you reach Shenfield
Common, and cross the line by what is locally known as
the Seven Arches. It is at first rather an uninteresting
road, but not for long, as a very little distance further on it
has woods on both sides of it, and a mile beyond dis-
closes a bit of common land on the left, literally choked

with masses of bracken and stunted trees of varied
species. This is at the "four cross ways." Up the road
to the left you see an entrance to Thorndon Park, straight
on leads to Little Warley, and the lane at the right, a
very pretty one, brings you almost immediately to
Warley Barracks. The barracks have no attraction for
the photographer, but the undulating and broken expanse
behind them is known as Warley Gap, an altogether
delightful common with gorse, broom, blue-bells and
bracken crowding together on its sandy banks, affording
at all times mutual contrasts. From the high ground
you look down its descending slopes, over the blackened
ruins of a recent fire, to the low lands of the Thames
valley, and see a distant gleam of river, with Tilbury
and Gravesend ten miles away, and the Kentish hills
beyond. At the foot of the "gap" stands a farm, and the
road curves suddenly round to the left by a small group
of cottages, but before passing these, look over the gate
through which runs a footpath to East Horndon, and you
will find a very charming hedge-bottom, with cattle studies
usually very near it. A little further up the road we reach
Warley Common, a very pretty expanse, with picturesque
cottages on its southern side, a pond and a flock of geese.
Leaving this behind, a rise in the ground brings us to
the top of a hill overlooking a splendid sheet of water
at the foot of Thorndon Park; fully half of it is grown
with rushes, and the woods of Thorndon make a very
effective background.

There are more cottages here—Childerditch parish—
and some of them have a very picturesque appearance,
with no lack of little children to aid in composition. The

tiny brook is crossed by a plank bridge; and a pathway, after crossing a couple of fields, enters the ground of Thorndon Park; more sheep and tree studies, together with a little stream and a bridge about half-way across the park. There are notices as to trespass here also, but Lord Petre is reputed to be willing to grant permission to wander from the beaten tracks if application is previously made by letter by intending visitors.

At the other end of the park is Herongate, and from here to Brentwood, through Ingrave, is three miles. East Horndon Station on the London, Tilbury and Southend line is near, and for this end of Thorndon Park and Warley Common more convenient for Londoners.

CHAPTER XXIII.

CONSTABLE'S COUNTRY.

Everybody knows that Constable's Country lies up the Stour valley, beginning, let us say, at Manningtree, and ending at Stoke by Nayland, roughly computed a distance of ten miles, and comprising, in the language of the land-agent, arable and pasture land, marsh and reed-grown dyke, upland and valley, together with that stretch of quiet river known as the Vale of Dedham.

There is nothing in the physical features of this country that may not be looked for and found in any other English landscape, although Constable was wont to say its beauty made him a painter. It is just the beauty conveyed by the existence of a running river, and had Constable been born on the Lea instead of the he would have idealised its scenery to an equal extent.

I say idealised of set purpose, for if the latter-day photographer expects to find Constable's Country as the painter left it, or rather *as he chose to see it*, there can only be one result—inevitable disappointment, for, to borrow a phrase from Hamerton's "Imagination in Landscape Painting," the painter has "far more the impulse to be imaginative than the anxiety to be accurate." Constable

K

had it very strongly marked, and perhaps it explains many things with regard to his works, throwing light on the unexpected appearance of recognised landmarks of the country-side—such as Dedham Church—in altogether impossible places.

A photographer has no such licence as this, for is not photography " a purely scientific or unfeeling art " (Mr. Hamerton again), which drawing can only exceptionally become " in the hands of a purely scientific or unfeeling artist," whatever that may mean?   The fact is, we are severely handicapped, and it is denied us to be able to write as Constable did to his engraver Lucas, " I long to see the church now that it is removed to a better place —two fields off." Alas! that a painter can take such a mean advantage of us.

It was not to question Constable's treatment of topographical subjects or his method of committing them to canvas I proposed to deal with in this paper; it was rather to identify—where identification is possible—one or two of the more famous of his pictures, with places that really are to be found in the Stour valley, and to see them with the aid of a photographic lens instead of the eye of imagination.   There is only one fault to be found with the Stour valley, and it is that the painter has been there before the photographer, yet those who can find nothing new in its varying phases lack discrimination, or, having found it without satisfaction, must be very exacting.

The Stour is presumably a navigable river ; presumably because I have followed it from its junction with the Orwell at Harwich to Manningtree, and from there right

up to Bures, where Constable's Country may be said
to terminate, and have found barges in nearly every reach.
Yet their appearance in places is puzzling, for there are
districts where no tow-path is visible, particularly between
Stratford St. Mary and Higham, and from there again to
Langham and Stoke Nayland, where the hedges come
right up to the water's brink, and no passage by land is
possible. Added to this, one is continually coming
across low wooden railings, set up over the towing-path
at about two feet from the ground. These last are not
insurmountable, and a passage in "Leslie's Life of
Constable" explains their *raison d'être*. The writer,
speaking of " A Canal Scene," exhibited in 1825, says,
·' The chief object in its foreground is a horse mounted
by a boy leaping one of the barriers which cross the
towing-paths along the Stour (for it is that river and
not a canal) to prevent the cattle from quitting their
bounds. As these bars are without gates, the horses.
which are of a much finer race, and kept in better condition than the wretched animals that tow the barges
near London, all are taught to leap." This might have
been written of the present time, and it is an extraordinary sight in the lower river below Flatford to
discover, coming through the flooded meadows and marsh
land, a barge making its way down the Stour with the
ebbing tide ; no river to be seen, no towing-path, only a
vast expanse of water, dotted about with pollarded
willows, by the side of which a horse, splashing belly
deep, is piloted by its boy rider. It seems to the un-
initiated that one false step would end in disaster, and yet
the scene holds one spell-bound in its striking picturesque-

ness; the line of willows marks the submerged path clearly enough, and this is closely followed. The Stour is open to the pedestrian almost everywhere, and in ordinary weather the tow-path may be followed in easy stages from Manningtree to Flatford Mill, thence to Dedham, Stratford St. Mary, Higham, Langham and Boxted, and inland to Stoke by Nayland, all of which places I will deal with presently.

The ordnance map for the whole district one inch to the mile, is No. 224, and I look upon it as an indispensable accompaniment to a photographic excursion in Constable's Country, because every bridle lane and field path is marked upon it, and in marshy land, where one is liable to be pulled up by dykes and impassable creeks, the utility of a trustworthy map is appreciated. Below Flatford it behoves one to be careful of the tide, for some parts, if not most, of the tow-path are submerged at high water, and to be caught by the tide midway between dry lands is fraught with inconvenience.

Dedham is the centre of Constable's Country, and it is usual to reach this village from Ardleigh, the station beyond Colchester. Yet Manningtree is better, if slighty further away, for the road from there is full of interest, and leads you past Flatford, the scene of so many of Constable's paintings, exhibited under various other titles. It is rather curious that Flatford is not designated by the ordnance map, because there is not the slightest doubt of its identity either as to name or situation. It lies between Manningtree and Dedham—by the river about half-way—and is marked on the map as "Valley Farm," &c., yet it is a cluster of houses, a little hamlet

by itself, and its association with Constable should have made it worthy of mention by its own proper name. His painting, "The River Stour," is Flatford. So also is the "Lock on the Stour," both being easily recognised by the little wooden bridge with the thatched cottage at the right and Dedham Church peeping out between the trees over the meadows in the distance, but they are Flatford idealised, and altogether outside the range of practical photography. If you look at the construction of Constable's lock, you will find that it is different to those found on the Stour, and is without the crossbars that invariably top them. These may be a later addition, but I do not think so, as they bear unmistakable evidences of antiquity, and from the first lock at Brantham Mills right away to the upper reaches of the river "their unanimity is wonderful." It will not do, of course, to compare a painter's latitude in composition, and his disregard of typographical accuracy with photographic representation of things as they are, but it is curious to notice in this particular picture of "The Lock" how Constable has fallen into an unnecessary breaking up of unity in order to introduce Dedham Church vignetted through the trees at the extreme left. The painter has shown us photographers, however, the value he put upon foregrounds and long shadows in composition, and the shimmer on the water in the quiet pool between the lock and the pool is altogether delightful.

Constable's skies, again, are marvellous, and it is remarkable what grand effects he was continually making by free use of the heavy *nimbus* and *cumulus* rain-clouds. Of his "Hadleigh Castle" he said himself that it was

"mighty fine, though it looked as if all the chimney sweeps in Christendom had been at work on it, and thrown their soot bags up in the air." It was Fuseli who complained that, although Constable was always picturesque, of fine colour, and had the light always in the right places, his paintings made him call for his great coat and umbrella. The great painter made especial study of skies, and his method of classifying them should commend itself to photographers, showing, as it does, that there was no haphazard selection when he came to adapt them to his landscapes. Leslie says : " Twenty of Constable's studies of skies made during the season are in my possession, and there is but one among them in which a vestige of land-scape is introduced. They are painted in oil on large sheets of thick paper, and all dated with the time of day, the direction of the wind, and other memoranda on their backs: "5th Sept., 1822, 10 o'clock morning, looking south-east, brisk wind at west, very bright, and fresh grey clouds running fast over a yellow bed about half-way in the sky. Very appropriate to the 'Coast at Osming-ton.' "

Tonality, a newly-created bugbear to photographers, never troubled Constable; and the painter, writing to his friend Fisher—Dean Fisher, I think it was—says : " That landscape painter who does not make his skies a very material part of his composition, neglects to avail himself of one of his greatest aids"—a precept he thoroughly carried into execution. This is a digression, and I have only quoted these extracts from an interesting work because they bear directly upon photographic methods. Perhaps their introduction will be pardoned.

One can spend a very pleasant day round and about
Flatford and the lanes and field paths that connect it
with East Bergholt and Dedham, the next village higher
up the stream, and the largest in the valley. When
Mr. Lyonel Clark's "Dedham Lock" was exhibited a
year or two ago, it created some sensation, and rightly
so. A painter remarked to me on the repetition or
imitation of Constable's treatment of sky and foliage he
noticed in it, and asked whether the latter was peculiar
to this part of Essex or Suffolk. It was conceded at the
time, I think, that brush work had been done upon the
photograph for effect, and whether legitimate or not,
which is altogether outside the question in the mention
of it here, the result attained was pleasing in the extreme.
But the picture is not a contemporary record, for it
would puzzle Mr. Clark or anyone else to find Dedham
Lock in the same mood ; the willows that had previously
escaped pollarding have at length suffered that indignity,
and those that were pollarded before have now been
pollarded the more. Alas ! for Dedham Lock, the ground
is strewn with lop, top, and branch, and the hideous
square white mill looks grim and uncompromising across
the river, the more so from its nakedness. The bridge is
still the same, and is thoroughly typical of the Stour,
being, like the generality of them, built of wood, and of
a long, rambling design. Over it runs the road to East
Bergholt, Constable's birthplace, a pretty little upland
parish about midway between Manningtree and Higham.

The Vale of Dedham, up to Stratford St. Mary, is
very beautiful—a quiet beauty—and full of possibilities
in the way of picture making. Generally speaking, it is

well timbered, and may be followed on either side of the
river.   Towards Stratford, on the Suffolk side, the
marshy land bears upon it some beds of coarse flowering
reeds and marsh grass that greatly assists the photographer in foreground composition,

Stratford St. Mary is a quaint little place of no great
size situated on the main road between Colchester and
Ipswich ; it has some pretty thatched cottages, a lock
and a backwater, but its most prominent feature is a
huge square-built mill, which until lately was dedicated
to the manufacturers of maccaroni, but for the present is
lying idle, and the water that comes rushing over the
weir spends its force in vain.   Beyond the lock, a pretty
one overhung with Scotch firs and ivy-grown elms, the
river generally has flooded the low-lying fields, and
pollarded willows—a special feature of tree life in Constable's country—are clumped together here and there, in
other places standing solitary, reflected in the still waters,
and across the flood red-roofed cottages, with the tower of
Stratford Church, combine to make a striking picture.

Looking at the map you will see that the river takes
a circuitous bend by way of Higham before it reaches
Langham, and though there is a pretty church at the
former village, it is better to join the river again at
Langham Mill by the road on account of the absence in
places of a reliable towing path.   This road branches off
from the higher ground at Stratford through Langham
Park, and is a charming one to follow ; it runs past the
prim little church on the crest of a hill overlooking the
river and the Glebe Farm below it, the scene of another
of Constable's paintings.   The latter-day pilgrim sees it

not with the painter's eyes, however, and he may search
in vain for the heavy foliage that comes crowding in at
the right of the picture, the brook trickling at the foot of
the slope, wherein a cow is drinking, for, as Leslie says,
"the rising ground and the trees on the right hand are
imaginary, as the ground in reality descends rather
sharply on that side of the church." The bubble of
illusion is pricked again for the matter-of-fact photographer, and his reflections—unwittingly, perhaps—take
a sardonic turn unwarranted by the occasion.

On the river again is Langham Mill, prettier than
any since leaving Flatford, but not of much account after
all for photographic purposes. The bridge is very
curious, and worth looking at seriously; and the lock
and backwater are the most charming on the Stour, on
account of the wildness of the vegetation to be found on
the little island separating the main stream from the
backwater and the background of trees that overhang
the river higher up. That luxuriant sedge is not easily
forgotten. "The Valley Farm"—the countryside abounds
in valley farms—between Langham and Boxted, the
next village up stream, is very picturesque indeed, with
its square, high chimney stacks and surrounding elms,
and it should have been mentioned before that the
easiest way to reach it is by a footpath on the South or
Essex side of the river which runs right through the
farm-yard on its way to Boxted. It is possible to take
the other side, but at times it is very bad travelling, and
a lot of time may be wasted in trying to find where the
tow-path has disappeared, and why a dyke must be
crossed in order to pursue the even tenor of one's way.

Boxted, again, has a mill, a weir, and backwater, and a long wooden bridge carrying over the road to Stoke by Nayland, lying about two miles away in a north-westerly direction, the terminus, as it were, of Constable's Country; the river trends first North, then due West, and has high, grass-grown banks like those of the sea wall round the Essex coast.

I can imagine no more delightful valley than that of the Stour on which to spend a long holiday; its very primitiveness and comparative seclusion will charm the lover of Nature, whether he be photographer, fisher or dreamer, or a combination of the three. There are no Bank Holiday crowds in the Vale of Dedham, and one may wander for miles along the banks of the Stour without coming across a human being. A desirable outlook for the photographer of landscape? I think so. It will be well to remember that Dedham is in the heart of Constable's Country, and from there one can easily reach any other village marked on the map, either by road or field path, for the authorities have been very free-handed with the latter, and they are plainly marked on the ordnance survey. Dedham, moreover, is almost the only village where reasonable accommodation is to be had; it has two hotels, and at the Marlborough Head, a rambling and old-fashioned inn, one can be very comfortable at such a moderate charge that I am afraid to state it.

Dedham, as I have said before, is usually reached from Ardleigh Station, and is about three miles from that place. From Manningtree it is at least four, and as it may be rather difficult to find the way without a map,

I will describe it briefly. The road turns under the railway bridge to the hamlet of Cattawade, crossing the Stour estuary by a long bridge, built on wooden piers, then it bears to the left again past Brantham Mill, and in two miles reaches East Bergholt. But before entering the village a lane to the left dips down to Flatford Bridge, and from there is a good gravel path by the river to Dedham. Another way, a nearer and better for photographic work, is by a footpath across the cornfields, immediately to the left after passing Brantham Mill, saving a mile in distance and rendering it possible to follow the lower reaches of the river, on which are to be found many pleasing pictures of a certain kind, flooded marsh, and "salting." This footpath joins issue with the aforesaid lane at Flatford Mill, and is easily traced,. but in winter or after heavy floods, I am compelled to admit it, the wayfarer will regret that he has not stuck to dry land and the road that goes by Bergholt.

Extract from Letter received March 13th, 1893:—

"I never use any other Plates than Edwards' Isochromatic."

(Signed)

BERNARD ALFIERI.

B. J. EDWARDS & Co.,

HACKNEY, LONDON.

INDEX.

W. B. WHITTINGHAM & Co., LTD., "Charterhouse Press," Charterhouse Square, E.C.

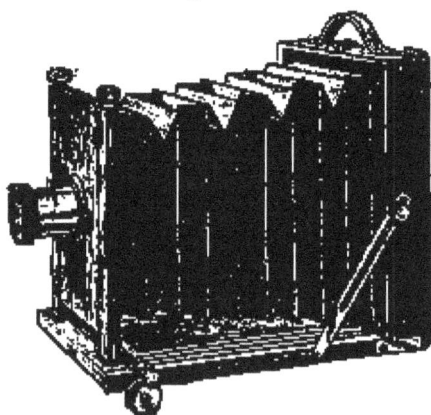

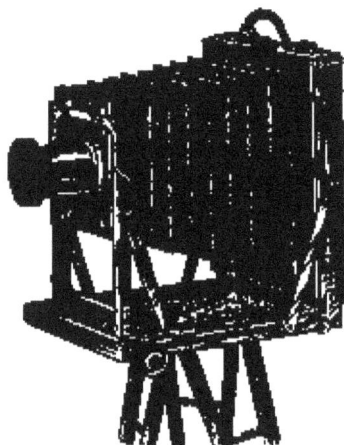

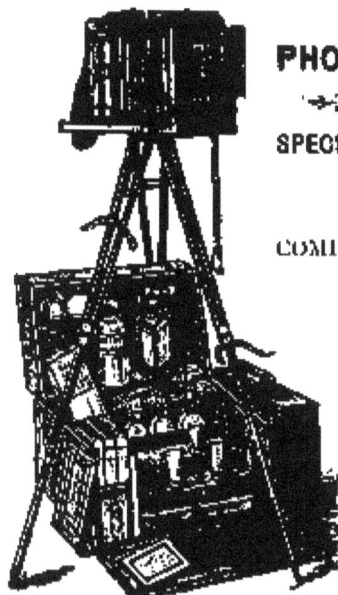

www.ingramcontent.com/pod-product-compliance
Lightning Source LLC
Chambersburg PA
CBHW021809190326
41518CB00007B/510